COMMUNICATIONS TECHNOLOGY EXPLAINED

About the Wiley-BT Series

The titles in the Wiley-BT Series are designed to provide clear, practical analysis of voice, image and data transmission technologies and systems, for telecommunications engineers working in the industry. New and forthcoming works in the series also cover the Internet, software systems, engineering and design.

Other titles in the Wiley-BT series:

Software Engineering Explained
Mark Norris and Peter Rigby, 1992, 210pp, 0-471-92950-6

Teleworking Explained
Mike Gray, Noel Hodson and Gil Gordon, 1993, 310pp, 0-471-93975-7

The Healthy Software Project
Mark Norris, Peter Rigby and Malcolm Payne, 1992, 210pp, 0-471-92950-6

High Capacity Optical Transmission Explained
Dave Spirit and Mike O'Mahony, 1995, 268pp, 0-471-95117-X

Exploiting the Internet
Andy Frost and Mark Norris, 1997, 200pp, 0-471-97113-8

Media Engineering
Steve West and Mark Norris, 1997, 250pp, 0-471-97287-8

ISDN Explained, Third Edition
John Griffiths, 1998, 306pp, 0-471-97905-8

Total Area Networking, Second Edition
John Atkins and Mark Norris, 1998, 326pp, ISBN 0-471-98464-7

Designing the Total Area Network
Mark Norris and Steve Pretty, 1999, 352pp, ISBN 0-471-85195-7

Broadband Signalling Explained
Dick Knight, 2000, 456pp, ISBN 0-471-97846-9

eBusiness Essentials
Mark Norris, Steve West and Kevin Gaughan, 2000, 320pp, ISBN 0-471-85203-1

Communications Technology Explained
Mark Norris, 2000, 396pp, ISBN 0-471-98625-9

COMMUNICATIONS TECHNOLOGY EXPLAINED

Mark Norris
Norwest Communications, UK

JOHN WILEY & SONS, LTD
Chichester • New York • Weinheim • Brisbane • Singapore • Toronto

Other Wiley Editorial Offices

John Wiley & Sons, Inc., 605 Third Avenue,
New York, NY 10158-0012, USA

WILEY-VCH Verlag GmbH
Pappelallee 3, D-69469 Weinheim, Germany

Jacaranda Wiley Ltd, 33 Park Road, Milton,
Queensland 4064, Australia

John Wiley & Sons (Asia) Pte Ltd, 2 Clementi Loop #02-01,
Jin Xing Distripark, Singapore 129809

John Wiley & Sons (Canada) Ltd, 22 Worcester Road,
Rexdale, Ontario M9W 1L1, Canada

Library of Congress Cataloging-in-Publication Data

Norris, Mark.
    Communications technology explained / Mark Norris.
      p.   cm.
    "Wiley-BT series"
    Includes bibliographical references and index.
    ISBN 0-471-98625-9 (alk. paper)
    1. Telecommunication.   I.   Title.
    TK5101.N784   2000
    621.382—dc21
                                              99-052680

British Library Cataloguing in Publication Data

A catalogue record for this book is available from the British Library

ISBN 0-471-98625-9

Typeset in 10/12pt Palatino by Footnote Graphics, Warminster, Wiltshire
Printed and bound in Great Britain by Antony Rowe Ltd, Chippenham, Wiltshire
This book is printed on acid-free paper responsibly manufactured from sustainable forestry, in which at least two trees are planted for each one used for paper production.

Contents

Foreword

Information networks are the backbone of modern business. Without them, the supply of money, fuel and food would rapidly diminish, civil order would be threatened, transportation would suffer and entertainment would revert to medieval times.

So it should be no surprise that the telecommunications and computing sectors have, for some time, been the most buoyant for profits, growth and jobs. To succeed in this upbeat and fast moving area requires a broad understanding of technology. And not just the latest set of buzzwords (although these are always useful) but a firm grasp of basic principles and their practical realisation.

This book provides exactly that—a clear explanation of communication technology. It has several unique features:

- it brings together all of the essential technology that underpins modern Information networks and places it in useful context. The aim is to provide a primer, based on hard fact, that prepares you to work in and with the telecommunications, network and computer industries;

- it concentrates on explaining why things are the way they are and presents as straightforward a picture as is possible. Considerable attention has been paid to making the topics covered accessible and to impart understanding, not just data.

It is intended for a wide range of readers:

- essential reading for anyone who needs to appreciate the breadth of communications technology. The book is particularly useful for those who find themselves (either by design or by circumstance) doing business with computing, network or telecommunications specialists;

- a valuable professional updating guide system designers, computer scientists, IT specialists, telecommunications engineers, system analysts and software designers, as well as business and information planners;

- a useful text for final year and postgraduate students in computer science, electrical engineering, IT and telecommunications courses.

ACKNOWLEDGEMENTS

The author would like to thank those kind individuals who contributed ideas, advice, words, pictures and even volunteered (I use this word in its loosest sense) to review early drafts. Some took a non-trivial amount of time out of their busy schedules, so deserve special mention: John Atkins, Professor David Bustard, Andrew Frost, Dr John Masterson, David Sutherland and Steve West. Many thanks—it is a better book for your knowledge, wisdom and observations.

And it would be remiss of me not to mention Ann-Marie Halligan, Sarah Lock and Laura Kempster at John Wiley who supported and encouraged me through the production of this book.

Mark Norris

About the Author

Mark Norris runs Norwest Communications, a specialist networks and computing consultancy. He has over 20 years experience in software development, computer networks and telecommunications systems and has managed dozens of projects to completion from the small to the multi-million pound, multi-site. Mark has worked for periods in Australia and Japan and has published widely over the last 10 years with a number of books on e-business, software engineering, computing, project and technology management, telecommunications and network technologies. He lectures on network and computing issues, has contributed to references such as Encarta, is a visiting professor at the University of Ulster and is a fellow of the IEE. He can be found at mnorris@iee.org.

1

The New Communications Business

According to the record of the last five years, the number of computers attached to networks is rising too fast to measure. Only by comprehending the full force of the computer juggernaut can one anticipate the future of the information age.

George Gilder

Few industries change as rapidly as telecommunications. Looking back as little as 10 years, this business could readily be characterised as a government owned voice network that evolved slowly and only after consensus had been reached across the industry. Most of the development effort that was poured into national and international networks was dedicated towards making an established service more reliable and more widely available.

Since then, dramatic change has come to the industry. A combination of forces—some technical and some organisational—have altered the face of the telecommunications business forever. Many of the established players have transformed themselves from network providers with an engineering focus into service providers with a customer focus. Along the way, a whole raft of new services have appeared, and in a very short space of time. To all intents and purposes, a brand new communications business has emerged.

This chapter aims to paint a broad picture of this new communications business. In particular, many of the basic concepts are explained, along with the way in which they are applied. Subsequent chapters will elaborate on some of these brief descriptions. For now, the whistle stop tour.

1.1 SOME BASIC CONCEPTS

Behind all of the changes in telecommunications, there are a number of key ideas and concepts that surface time after time and need to be understood. Those that are relevant to the issues in this chapter are as follows.

Analogue and digital networks

Early telecommunications were analogue. They used continuously variable signals to convey information. The quality of speech across these analogue networks was determined by the amount of the speech spectrum that could be carried. Just over 3 kHz was accepted as a reasonable compromise of cost and quality. The 300 Hz to 3.4 kHz range (with a 4 kHz peak) formed the basis of all analogue telephone networks. As the telephone network moved from analogue to digital technology, the established voice signals were transformed into 64 kbps bit stream, as illustrated in Figure 1.1.

Computer communications, which are invariably based on discrete, digital signals, can use analogue connections but are limited by the available bandwidth (as explained in the section on bandwidth that follows).

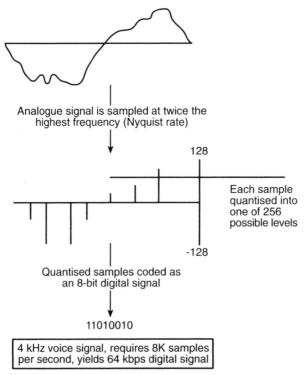

Figure 1.1 From Analogue and Digital

Recently introduced all-digital networks are ideally suited to computer networking and the capacity of these networks has grown very rapidly (e.g. from 64 kbps in the mid 1980s to 10 Mbps+ by the mid 1990s) and they can carry a mix of voice, data, text and pictures.

Circuit-switching and packet-switching

The distinguishing feature of circuit-switching is that an end-to-end connection is set up between the communicating parties, and is maintained until the communication is complete. The public switched telephone network (PSTN) is a familiar example of a circuit-switched network. Indeed, it is so familiar that many people are not aware that there are other ways of doing things (Figure 1.2).

Communication between computers, or between computers and terminals, always involves the transfer of data in blocks rather than continuous data streams. Packet switching exploits the fact that data blocks can be transferred between terminals without setting up an end-to-end connection through the network. Instead they are transmitted on a link-by-link basis, being stored temporarily at each switch *en route* where they queue for transmission on an appropriate outgoing link. Routing decisions are based on addressing information contained in a header appended to the front of each data block. The term packet refers to the header plus data block (Figure 1.3).

One further distinction that should be made is between connectionless and connection-orientated services. With the latter, an end-to-end link is established for the duration of the call. The familiar telephone call is connection-orientated, as are some packet data calls. Connectionless on the other hand assumes that there is no set connection—every packet is left to find its own way from source to destination.

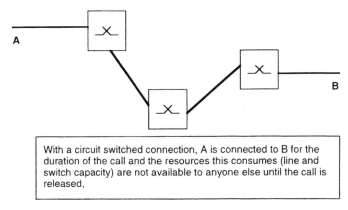

A

B

With a circuit switched connection, A is connected to B for the duration of the call and the resources this consumes (line and switch capacity) are not available to anyone else until the call is released,

Figure 1.2 Circuit switching

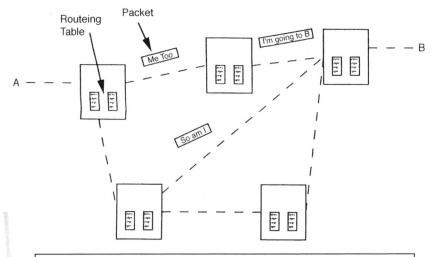

There is no concept of a call in a packet switched nework.
Each packet waits for the resources it needs to become available,
so there is prospect of delay (but not of blocking) and it is likely
that different packets will take different routes (and hence experience
different delays). It is left to A and B to ensure that the right packets
are received in the right order.

Figure 1.3 Packet switching

Congestion and Blocking

In a packet-switched network, packets compete dynamically for the network's resources (buffer storage, processing power, transmission capacity). A switch accepts a packet from a terminal largely in ignorance of what resources the network will have available to handle it. There is always the possibility therefore that a network will admit more traffic than it can actually carry with a corresponding degradation in service. Controls are therefore needed to ensure that such congestion does not arise too often and that the network recovers gracefully when it does.

In a circuit-switched network the competition for resources takes the form of blocking. This means that one user's call may prevent another user from gaining access. Since the circuit is reserved by the user—irrespective of what they send—for the duration of their call, no-one else has any form of access until the call is cleared. Traditional circuit-switched networks are designed to balance the amount of equipment deployed against a reasonable level of access for the users of that network.

Performance

A circuit-switched network, such as the PSTN, provides end-to-end connections on demand, provided the necessary network resources are available. If there is no available path, the call attempt is blocked and you have to try again later. Once connected, the end-to-end delay is usually small and always constant. Furthermore, other users cannot interfere with the quality of communication.

In contrast, with a packet-switched network, you will not be blocked but you may be delayed. Packets travel to their destination one hop at a time and have to queue for transmission capacity at each switch en route. The cross-network delay is therefore variable, depending on the volume of traffic encountered. In addition, the effective bandwidth between two points cannot be guaranteed.

Multiplexing

In order to use available capacity as efficiently as possible, many logical connections are often carried over one physical connection. The basic mechanism by which this is achieved is known as multiplexing. There are two forms of multiplexing—time or frequency division. In the latter, each connection is allocated a time slot in a digital stream; in the latter each has a frequency allocation in a broadband analogue link (Figure 1.4).

In the public telephone network, there is a hierarchy of multiplex links. This used to be implemented with carrier systems that put 12 speech

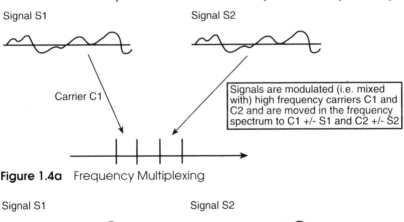

Signal S1

Signal S2

Carrier C1

Signals are modulated (i.e. mixed with) high frequency carriers C1 and C2 and are moved in the frequency spectrum to C1 +/- S1 and C2 +/- S2

Figure 1.4a   Frequency Multiplexing

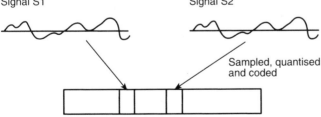

Signal S1

Signal S2

Sampled, quantised and coded

Both signals are digitised (e.g. turned from analogue voice signals into 64 kbps data streams) and allocated a time slot in a high speed digital stream

Figure 1.4b   Time Multiplexing

channels into a group carried between 64 kHz and 108 kHz. These groups were, in turn, aggregated into 60 channel supergroups between 312 kHz and 552 kHz.

More recently, bulk channels have been carried within the telephone network via a multiplex arrangement known as Plesiochronous Digital Hierarchy (PDH) (Figure 1.5). This treats each telephone circuit as a 64 kbps digital channel and (in Europe) aggregates 30 channels into a 2 Mbps stream. Two additional channels are added for synchronisation and signalling. Higher order multiplex links are used between transit exchanges at 8 Mbps (120 channels), 34 Mbps (480 channels) and 140 Mbps (1920) channels. The hierarchy is slightly different in the US and Japan (see lower half of Figure 1.5).

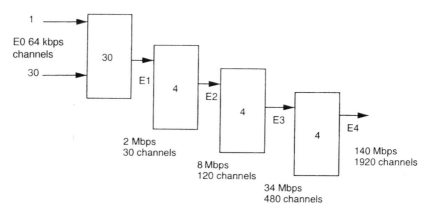

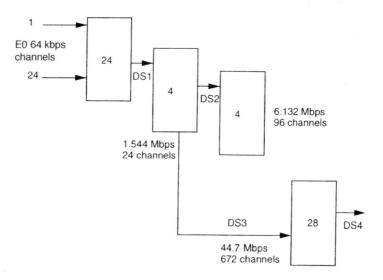

Figure 1.5 The Plesiochronous Digital Hierarchy

Data Rate	STS (SONET Electrical Level)	OC (Optical Leve)	STM (SDH Electrical Level)
51.84	STS-1	OC-1	
155.52	STS-3	OC-3	STM-1
622.08	STS-12	OC-12	STM-4
1244.16	STS-24	OC-24	
2488.32	STS-48	OC-48	STM-16
9953.28	STS-192	OC-192	STM-64

Figure 1.6   The Synchronous Digital Hierarchy

Typically, the higher order multiplex links employ optical links whilst the lower levels are copper.

The current drive within the network is to move from the PDH to a Synchronous Digital Hierarchy (SDH) (Figure 1.6). SDH, which is known as Sonet in the USA, is an international standard that standardises transmission around a rate of 51.84 Mbps (known as STS-1). Multiple rates of this basic rate comprise higher rate streams: STS-3 at three times STS-1 etc. As well as operating at rates more in line with modern data rates, SDH allows a single channel within a multiplex stream to be accessed (not possible with PDH).

Signalling

With the increase in complexity and flexibility of modern communications, signalling is becoming ever more important. Early signalling systems tended to be carried in the same space as the signal that was being carried, thus limiting control to pre- and post-connect. Modern telecommunications signalling systems carry their information (i.e. their call control data) separately from their payload (i.e. the user data) and can thus effect control during the course of a call.

This is particularly important for setting up network features and a standard system called CCITT Signalling Systems No. 7 (often referred to as C7) has been developed for this purpose. C7 is one of the basic enablers of the Intelligent Network.

In packet-switched networks, signalling is an integral part of each packet. Typically, a packet header will be defined and this will contain items such as destination address, the packet sequence number and priority.

Bandwidth

The bandwidth of any communication channel is the major determinant of its information carrying capacity. The precise definition is the range of frequencies that the channel is capable of transmitting.

A standard telephone channel can support frequencies in the range 300 Hz to 3.4 kHz and it is this bandwidth that sets a limit on the rate at which digital information can be carried over the link. Claude Shannon discovered the exact relationship and devised the formula:

$$\text{Maximum bit rate} = w \log_2(S/N + 1)$$

where w is the bandwidth (Hz), S the signal strength and N the noise strength. Shannon's formula shows that signal-to-noise ratio has significant impact on the amount of information that can be transmitted down a wire. Hence, a 3 kHz channel with typical S/N ratio of 30 db permits a little less than 30 kbps. This may look a little odd, as standard voice band modems can operate at speeds up to about 60 kbps The apparent contradiction can be explained by the fact that a change of state can represent more than one bit of information. In much the same way ISDN modems can (using data compression) reach about 115 kbps.

It is possible to carry more information over local telephone lines using techniques such as Asymmetric Digital Subscriber Line or ADSL. With ADSL, higher frequencies (4–900 kHz) are used. The bandwidth is divided with each division carrying a part of the overall data transmitted. The penalty for operating at these higher frequencies is that reach is limited (a distance of 3 km is typical at 2 Mbps)

We will return to these topics in greater detail later on in the context of their practical application.

1.2   TECHNOLOGICAL CHANGE

It is becoming increasingly difficult to differentiate telecommunications from computing. The increasing use of computers for both switching and network management has meant that the major technical challenge before most telecommunications companies is, in essence, the effective deployment of networks of communicating computers. At the same time, the computing industry has been building computers that have ever more sophisticated communication capability. Their aim is to promote easier and more natural communication between computers.

Convergence
The convergence of computing and telecommunications has been under way for some time and, although far from complete, is well established. Many of the ideas born in the computing community (such as the Internet and distributed computing) are now central to most telecommunications companies' repertoire.

Less well established is the way in which consumer electronics and information providers will impact on telecommunications. With a number of joint ventures and alliances already established between recognised telcos (for instance, between BT and BSkyB), there will inevitably be an exchange of ideas and practices.

The key point being made here is that the breadth of 'telecommunications technology', that was once centred on voice switching and transmission, will become much greater.

Capacity
With computers playing a greater part in the communications business, there is a demand for more and more network capacity. There are a number of reasons why this is so, the main ones being:

- The once simple and few transfers between computers have become much more complex and commonplace. All of this is driven, to a large extent, by the fact that the usefulness of a network increases as more people come on line. This is the basis of Metcalfe's law, which observes that the value of a network is vested in shared information.

- Rather than simple text files, much of the traffic now being generated is a mix of text, video and voice. This is inherently hungry for bandwidth (e.g. a picture can easily be 100 kbytes, a video clip 100 Mbytes).

- Many computer users think little of attaching large files and distributing them to long distribution lists. Once, a network user could do little more than monopolise a single phone line. They can now generate enormous network traffic with ease .

Integration
In the working environment the Local Area Network (LAN) has become familiar to many people (Figure 1.7). Since the Ethernet was popularised in the early 1980s, it has been the predominant means of sharing information and resources in the office, the factory and the laboratory. Despite the advent of higher speed LANs (e.g. FDDI based), there have been few fundamental changes in local networking.

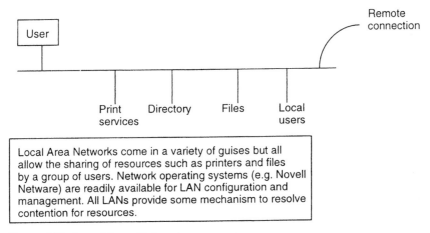

Figure 1.7 Local Area Networks

Asynchronous Transfer Mode (ATM) may change this. This promises to enable the transmission of voice, video, image and data across the same communication link and therefore allow the true integration and management of these services.

Since ATM is becoming an important standard for wide area networks as well as local ones, there will, as a consequence of this, be a coming together of local and wide area networks. This lays the foundation for globally distributed multiservice networks and is important for distributed organisations such as multinationals, banks etc.

Mobility

The cellular phone has rapidly become a fact of public life. It is now being developed to extend its usefulness through integration with pagers, computer, fax machines etc.

Cellular technology is moving from analogue to digital to make better use of available bandwidth. This also opens the way for the wireless Internet, as many data services can now make efficient use of cellular networks.

There will be an increasing use of wireless technology in cars and trucks, especially to derive location information from the Global Positioning System (GPS). In order to get the necessary global coverage there is likely to be a movement towards satellite transmission.

Local area networks in buildings will also use wireless technology thus overcoming restrictions in building structures and portability problems. Wireless access technology such the Infrared Data Association (IrDA) protocol and short-range radio link (in the unrestricted 2.45 GHz ISM 'free band' radio frequencies) can provide for rapid *ad hoc* automatic connections between devices.

Moving on from some of technological changes that are impacting the established telecommunications industry, there are some specific and continuing trends that impinge on every communications company. Before this, a brief overview is given of the where our starting place on this journey—the traditional telecommunications network.

1.3   THE PUBLIC SWITCHED TELEPHONE NETWORK

The telephone network that most people are familiar with has evolved over a number of years. In essence, it consists of a number of linked parts, the main ones being:

An *access layer* that is used to deliver the service to the end customer, and is currently primarily provided by twisted pair copper. There is some use of optical fibre for higher volume business customers, and coaxial cable is used for TV distribution. There is some use of multiplexing in the

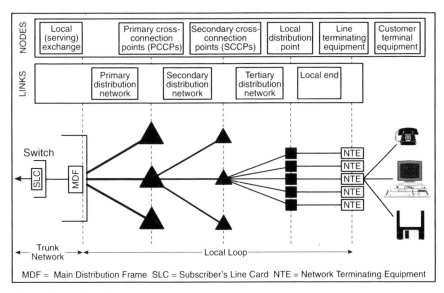

Figure 1.8 The Access Layer

access layer, but the primary interface to the switching and core transmission layers is through a main distribution frame (MDF) where physical cross connections are made between copper pairs. The overall hierarchy between the MDF and the network termination is shown in Figure 1.8.

A *core transmission layer* of the network that is provided primarily by digital transmission systems, known as Plesiochronous Digital Hierarchy (PDH), which are based on standards developed initially for the carriage of digital voice. This has an inherently inflexible multiplexing system, and is not optimised for flexible partitioning of bandwidth between the various classes of traffic that will be generated by the emerging applications seem likely to dominate bandwidth utilisation in the coming decades.

A *switch layer* that is used to provide the various switched services available, which are currently limited to a maximum bit rate of 64 kbps (the digital equivalent of one speech channel). There are three separate major switching services, circuit-switched PSTN and ISDN, which are partially integrated, and the packet-switched PDN. The switch layer is also used to provide cross connect functions for some leased services (Figure 1.9).

Controlling all this are various *network management* systems—usually one for each of the different types of equipment in the network. In addition, various service management systems are in place to control the different services provided. Service provisioning process is still dominated by the need to make physical interconnections at various points in the network, which makes an automated end-to-end process for service

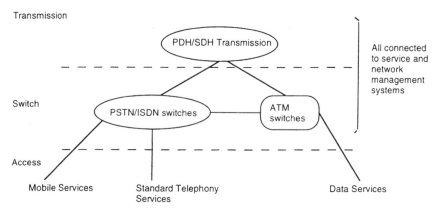

Figure 1.9   The Switch Layer in context

provision and rearrangement slow to effect and difficult to achieve with any measure of consistency.

Evolution of the PSTN
For the access layer, the primary focus of network evolution over the next decade will be on increasing the efficiency with which the existing copper pair distribution system can be used, while expanding capacity and flexibility through the introduction of optical fibre technology.

There will also be increased use of multiple technologies for service delivery as the viable technology options multiply and competitive pressures demand ever more cost-effective delivery. However, copper seems likely to remain the dominant form of delivery to the wall socket in domestic premises at least well into the first decade of the next century. The early introduction of ADSL and HDSL technology will allow the delivery of wider bandwidth services over the existing copper pair infrastructure. This will be supplemented by increasing use of point-to-point radio links for delivery of 2 Mbps and 34 Mbps services. The extent to which existing assets (i.e. copper pairs) can be made to 'sweat' will determine the rate at which fibre penetrates this market sector.

For core transmission, the current network is based on a hierarchy of time division multiplexing equipment, known as the PDH. This system lacks sufficient flexibility to deal with the demands now being placed on the configuration, operation and management of networks, and as a result the telecommunications standards body, the ITU, has defined a newer more flexible digital multiplexing scheme, referred to as SDH.

The SDH system takes advantage of more recently developed technology to provide superior inbuilt management and monitoring capabilities, simpler dynamic reconfiguration in response to changing demands or network failures, and higher reliability through a reduction in the number of separate items of equipment typically needed to provide the multiplexing function. SDH also provides a greater maximum link bandwidth (2.5 Gbps as against 140 Mbps) and this economically satisfies the emerging demand for wideband services.

Connecting a user to the service they require has always been one of the main challenges for the designers of wide area networks. Traditionally, switching has been the slowest and/or least flexible part of such networks. The problem is not one that is shared by local networks. Established protocols, such as CSMA/CD and token ring, allow a high capacity medium to be shared among a large number of users. In this latter case, connecting (for instance) a client to a server has not been a real issue. However, as the facilities and expectations of current LANs grow such that the traditional concerns of the WAN designer take hold, so switching technology will take a more central role.

SMDS and ATM based switching systems have been deployed to provide wideband services for business customers and are being developed to give similar facilities to domestic consumers.

1.4 THE INTERNET

Like the Public Switched Telephone Network, the Internet is a communications network that spans the globe (Figure 1.10). Unlike the PSTN, the Internet relies on packet-switching, rather than circuit-switching, was originally conceived as a data network rather than a voice network and has grown through the interconnection of many local area and regional networks, rather than being built as a single entity.

In essence, the Internet works by passing data using a standard communications protocol known as IP (Internet Protocol). The IP protocol can be used on many types of computer and over almost any network infrastructure: local and wide area. The Internet has an associated naming and addressing scheme, Domain Name Service (DNS) which allows resources (information, services and people) on the Internet to be easily located. It is

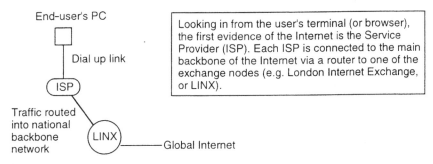

Figure 1.10 The Internet

the universal acceptance and wide availability of IP, DNS and some other key standards (such as SMTP for mail, TCP for end-to-end session control and HTTP for information transfer) that give the Internet its global reach, its broad availability and its vast user base.

One of the basic functions for which the Internet was built was to allow the transfer of information from one computer to another—and hence, from one person to another. This was initially effected thanks to a file transfer application (known as FTP) which was typically part of the TCP/IP package. Many networked PCs have a pre-installed FTP client program and simple FTP clients can connect to FTP hosts, view directories and download them according to user choice.

Typically, FTP servers require a username and password before they will allow connection. There are some FTP servers that will allow anonymous login; these require the username anonymous and a password that is either your Internet Protocol (IP) address or your Internet email address.

Perhaps the best known application associated with the Internet is the World Wide Web (WWW or web). The World Wide Web is an extension of the global Internet and builds on its established file transfer capability. Like FTP, web browsers (e.g. Netscape Navigator, Internet Explorer— software packages that give a user access the system) use a straight-forward file transfer protocol, known as HTTP, to connect to web hosts and can download information in any file format. The user is provided with a simple and intuitive interface for navigating information that is dis-tributed throughout the Internet. Access to Web pages does not usually require you to enter a username or password (although some commercial sites do require registration and/or a fee).

What makes the Web so much better than its predecessors is that the information, its addressing and referencing all abide by a common encoding standard, HyperText Markup Language (HTML).

What the Web browser does is to access and read files written in HTML. It is the HTML coding that defines the files' text layout and style when displayed on the browser. The HTML standard is quite basic but has a feature that fits the problems of distributed information well—the hyper-text link (or anchor). The hypertext link is a simple definition within the language that can be used to highlight a word or phrase within the text of the file. This is known as the anchor. This definition associates an address (a logical location) and a file reference to that anchor so that when it is selected by the browser an instruction is sent to fetch the file that is referenced. This is a bit of a long-winded explanation for a simple concept that most people are now familiar with and that works very intuitively in practise!

The power of hypertext used in this way has been demonstrated over the last few years by the phenomenal growth in the Web. The simple concept of linking one file to another, through an explicit and contextual reference in the text, means that many of the problems associated with redundant information and searching for a needle in a haystack can begin to be addressed. It also means that the Internet itself is no longer in the realm of the computer scientist or engineer. The Web has brought point-and-click usability to the Internet, just at a time when the growth in the home PC market has itself entered a period of vertical acceleration and the image of the PC in many a business is no longer the executive's toy but an indispensable tool for the job.

The World Wide Web
To reiterate an important (but often misunderstood) point, the World Wide Web is a multi-media. hypertext information retrieval system that sits on top of the Internet. The information stored in Web servers consists of text, images, sounds, and movies. A single Web document may contain all of these It draws on many of the Internet's concepts and combines them to provide the ultimate in point-and-click navigation systems—huge, ubiquitous, global and intuitive.

The Web is actually implemented as a client/server system. This means that software is distributed so that one end (the client) is usually the information consumer and the other end (the server) is the provider. A practical implication of client/server systems is that there are typically many more clients than servers. This imbalance is not a problem as the client to server interface is uniform, so there is an engineering task in achieving the right performance and client to server ratios.

The realisation of the client is the now ubiquitous Web browser. These are rich (and getting richer) in functionality and, as mentioned above, use a simple file transfer protocol (HyperText Transfer Protocol, HTTP) to connect to web servers to download and view files that have been written in HTML format. So the Web provides the user with a uniform and intuitive interface for navigating information that is distributed through-out the Internet.

The files (or pages) that a Web browser looks at appear within the browser window on your PC screen as formatted text and graphics. Embedded within the text of each document there are usually a number of hyperlinks. Each of these is a reference to another file somewhere on the Web. This reference is known as a Uniform Resource Locator (URL) and has a similar format to established Internet addresses (e.g. www.disney.com).

The real beauty of the Web is that its addresses do not have to be exposed directly to the user—the thinking is that this gets in the way of the Web's intuitive interface (but see below for a brief expose on IP addressing. Many people memorise URLs—they are usually pretty obvious and are routinely flashed up during TV adverts. Even if they cannot they could still easily navigate the Web. By clicking on a hyperlink with the mouse, the web client will retrieve and display the referenced document. The user can then browse through the Web of information by clicking on the links on each page.

The development of the PC is now at a stage where it is the vehicle for the delivery of multi-media information, whether it be text, graphics, video or audio. These media forms can now all be digitised and stored as data on computer disks at relatively low cost. The power behind the Web is in exploiting these trends through the very simple language of HTML and the concept of hypertext. HTML is used to define the layout of a computer screen combining text (in a range of typographical styles) and images whilst combining the ability to address other forms of information through hypertext links.

At the heart of hypertext is the anchor. This is a very simple construct within the HTML specification that identifies text, a single word or a string, as a hypertext link. An anchor is typically of the form:

<A HREF="http://www.expl.com/expl.html">Mark</A>

Where the link, Mark, would be shown by the browser in blue or some other colour rather than the default black. To click on the anchor would result in a request being sent from the client to the server (http://www.expl.com/) defined in the Hypertext reference (HREF). The request would be sent as an HTTP request, asking for the file expl.html. The file would then be sent to the browser and displayed.

If a file is not an HTML document, it will usually have a suffix that denotes its type (such as .txt, .wav, .mov, .ppt etc.). When the server encounters this suffix it will prepare the file according to its local type definition within its configuration and send it to the browser. The browser, on encountering a file type that is not HTML, will also look up the file suffix in its helper application configuration. If it is associated with a particular application, it will either save the file to disk or boot up the application so that it (and not the browser) can handle the file. For example, a file with suffix.pdf will result in the browser booting Adobe Acrobat for it to be read. Pdf stands for Portable Document format. It is a standard for document presentation that helps to ensure that a page of information, when viewed, is in the format that the originator intended.

HTML is easy to read and write since it consists of plain text delimited by tags (the special characters in chevrons) and, for this reason, it is easy to see just how extensible this language is. For example, an anchor always starts with <A ... > and ends in </A>.

To give some idea of its pace of development, HTML has seen at least two revisions per annum since it went under change control and now includes a whole raft of features, from tables to text fields, buttons, check boxes and context-sensitive image definitions. The next generation mark-up language, XML, is seen by many as the key to electronic business and other commercial applications of the Web as it allows standard types of information within a document to be identified.

The combination of features and concepts has pushed the Web far beyond the capabilities of the basic Internet utilities of mail, file transfer and bulletin boards. Much of this functionality has been subsumed in many implementations of the Web browser. This is just as well since the Web performs the functions that basic file transfer could not—the display of information within the files that were downloaded.

The more sophisticated browsers now bundle in mail and news client software within the same interface, offering a single software entry to all the resources on the Internet. This trend continues but in a more cooperative way, where Web browsers are now the platform for integration of plug-in functionality for specific purposes.

Because the basic concept of the Web—hypertext—is not dependent on file type, the linking action can be used by the requested Web server to interface to an application prior to sending any results back to the requesting client. What this means is that an anchor reference may be an application process that can accept input and return a result. For example, a typical Web page can be used to display a field and a button; the user can fill-in the field and, by clicking on the button, invoke a hypertext link that sends the filled-in field back to the server for the attention of an application script.

The script can be accessed by the server through a Common Gateway Interface (CGI) and so these scripts are generally referred to as CGIs. The CGI performs some operation on the text submitted through the filled-in field and returns a result back to the server/end user. This result would be in the form of a page of HTML. What has just been described here is, in effect, the front-end of a search engine for the Web.

Another benefit from hypertext is the ability to pull-in and collect a much greater body of information than hitherto. Special programs can be written to behave as Web browsers that can download a Web page, cache its links and then traverse all the links, pulling in each page, caching its links and so on. By crawling across the Web like this, the cache of links soon grows and forms a substantial database of information. Coupled with some software to perform searching on this database and the Web form (discussed above) it is relatively easy to build powerful Web search engines.

Because of the ease in which this can be done and the flexibility of HTML and of CGIs, it has not taken long for a multiplicity of different search engines to appear on the Web. Whilst it is sometimes difficult to choose which one to use, the Web search engines are still many times more powerful than previous Internet offerings such as WAIS. Some of the better known of the Web search engines are YAHOO, Lycos and Alta Vista.

The Internet Protocol
Internet addressing is based on the Internet Protocol (IP). This protocol, like any other, defines the expected behaviour between entities, a set of rules to which the communicating parties comply. IP is a network layer

protocol that offers a connectionless service, that is, it does not establish a link between calling and called parties for the duration of a call. It is higher level protocols such as the Transmission Control Protocol (or TCP) that sit on top of IP that add this (and, with it, assurances of delivery).

Because of its frequent association with other protocols, the Internet Protocol is often referred to as TCP/IP. This is the collective term used for its many derivatives implemented on personal computers and UNIX workstations. In practice Internet applications use either TCP or UDP (User Datagram Protocol); the differences between these two are that UDP is for unreliable (but fast) connectionless packet delivery, and TCP is for reliable connection-orientated byte-stream delivery. So UDP tends to be used for simpler services that can benefit from its speed, and TCP for more demanding ones where it is worth trading a little efficiency for greater reliability.

From an engineer's point of view, one of the real beauties of Internet lies in the way in which the IP datagrams are addressed to named hosts on the network. Each host on the Internet has a unique number, known as an IP address, and this is carried in the header of each packet sent (Figure 1.11).

The IP address is a 32 bit number that can be used to address a specific

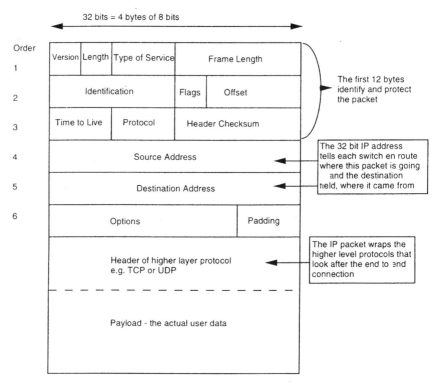

Figure 1.11 The IP packet header

network and host attached to the Internet. This does not mean that every Internet user has a permanent IP address. For instance, dial-in users using the point-to-point protocol (PPP) to connect to an ISP are 'loaned' one for the duration of their connection. Enduring IP addresses relate to a specific host on a specific network.

There is a structure to the IP address that allows both a network part and a host part to be defined. This split occurs in different positions within the number according to the class of the address, and there are four classes in common use. The first, Class A, uses just the first 8 bits to define the network and other 24 to identify hosts. There are, therefore, not that many Class A networks, but the ones that there are can support a lot of hosts. Classes B and C are more common but support fewer hosts and Class D is reserved for special purposes.

Resource location
To complete this section, we take a Uniform Resource Locator (URL) and break it down into its constituent parts, each of which has been explained at some point already.

Let us take the full version of the URL used earlier to explain the concept of an anchor:

$$\text{http://www.expl.com/expl.html}$$

and it can be broken down into a number of parts.

http://
This part tells the client which protocol it needs to use on this occasion. In this example, it is a standard access to a Web page, so HTTP is chosen. If the URL was targeted on a filestore, the first would most likely be ftp://.

www.expl.com
This is the readable version of the IP address as explained above. It resolves into the right address of the machine that you want to get your information from. So, assuming that the target machine is part of class B network, the Domain Name Service would resolve the mnemonic for the intended server into something like 146.139.16.17– the server's IP address. Sometimes, there are qualifications to this address field (e.g. www.expl.com:8080). The part after the semicolon is not part of the address as such—it is there to route an incoming request to a particular port on the host machine. This may be included for security or operational reasons.

expl.html
And finally, we get to the file that you are looking for. The extension .html should give you a warm feeling that all is well and that you are going to get an HTML coded document, as you would hope, using

HTTP. Many addresses are a lot longer than the one in our example, e.g. http://www.expl.com/net/faq/basics/expl.html. The intermediate references between the machine address and the final document reference are simply there to get to the right part of the directory structure on the host.

More detail on each of the elements of Internet technology can be found in a series of articles known as RFCs. This series may be known as Requests for Comment, but in practice it contains the definitive versions of all of the Internet's protocols.

1.5 DISTRIBUTED COMPUTING

In recent times, the ideas of teleworking, outsourced development and virtual teams have come very much to the fore in many organisations. They have been made viable through a combination of computing and telecommunications and rely heavily on the practical techniques for the operation of distributed information systems and on the construction of applications that run over them. Hence, just as the telecommunications industry has embraced computing as an integral part of its business, so the computing industry has sought to distribute its wares by adding communications capabilities to them.

There is little doubt that it is more complex to build effective information systems when the processing capability and data is scattered across various sites. It therefore follows that there must be significant advantage in doing so. The key benefit that accrues from distribution is flexibility. From the designer's point of view, there are more components and interactions to deal with but also more ways of completing a task without relying on a central database or mainframe computer.

The user benefit from this move away from a centralised and hierarchical structure and on to peer-to-peer interaction is that new ways of working become possible. For instance, collective design, where a group of experts, gathered for their specific knowledge rather than their being close at hand, contribute to a piece of work. The record of their combined thoughts can be held, in some format, on a remote machine but also made available for all to access and contribute to. Review, edit and formatting of the document is carried out during production and any ideas or suggestions are loaded onto a bulletin board so that all involved can see what is going on.

In a similar vein, it become possible to do off-site servicing, where the performance measures of a remote piece of equipment are collected for an operator to effect required changes. This sort of application is already in place in some specialist areas. For instance, the adjustment of turbines when they first go into service (and for ongoing maintenance and repair) can readily be carried out without the physical presence of an engineer. Files containing operating data (one way) and adjustments (the other way) do the job, with one of the best known distributed systems—the Internet—providing the means to do so.

As well as being able to work in a different way, there are a number of characteristics of information systems that change. A number of golden rules need to be taken on board by anyone who works in, wants to capitalise on or builds a distributed environment. There is a fundamental change of paradigm for many people accustomed to a more traditional environment and some of the resultant effects are explained in Table 1.1.

Table 1.1

In a Centralised System, it is . . .	Why the difference?	In a Distributed System, it is . . .
Local	In a distributed system some things will not be local—simply by definition. Parts of an application may, literally, be running on the other side of the world.	Remote
Sequential execution	Distribution implies interconnection of constituent parts. With many parts contributing to each users needs, applications tend to run in parallel.	Parallel execution
Synchronous	Most local interactions are serial and synchronous. Distributed systems introduce a much greater degree of asynchronicity and unpredictability.	Asynchronous
Centralised control	Distributed systems may encompass several autonomous sub-systems. No central control may be desirable or, indeed, possible.	Decentralised control
Total success or failure	Distributed systems can continue to operate if some parts fail or are unavailable. In a large distributed system (such as the Internet) it is unlikely that all parts will ever be fully available.	Partial success or failure
Fixed location	Components of a distributed system cannot be bound to a fixed location. They may be relocated from time to time—even while in use.	Changing location
Static configuration	Aspects of system configuration other than component location may also change whilst the system is in use. For example, duplicates of components may be introduced to share processing tasks.	Dynamic configuration

Table 1.1 *continued*

In a Centralised System, it is . . .	Why the difference?	In a Distributed System, it is . . .
Homogeneity	It is most unlikely that all of the components of a distributed system employ the same technology, or that those technologies will not change over time.	Heterogeneity
Unification	As systems grow or need to involve independent parties an imposed single structure becomes unworkable. Collaboration, cooperation and coexistence are essential.	Federation
Private	Many things that were private can become exposed to public view in a distributed system. Security has to be explicitly built in.	Public
Shared state	In a single, centralised system the current state of the system can be shared and known by all of the components. In the distributed world such shared state can no longer be assumed as there is no equivalent of shared memory space.	Local state
Single design	A large, dynamic distributed system usually evolves through organic growth, rather than as the result of a concerted design. Components need to be designed to be used, abused and re-used in such an environment.	Cooperating components

Each of these reversals from centralised thinking brings a degree of increased complexity along with a set of attendant difficulties, and none of these problems are entirely unfamiliar to programmers. However, many of them have been the provinces of specialists within particular application niches.

In principle, distributed computing ought to be easy. After all, it is only a case of tying together computers with networks. The industry as a whole has become very good at building fast, cheap, well understood, commodity computers. It has also been creating successful communications networks for decades. Furthermore, it is known how to make computers talk to each other over these networks.

The reason that distributed systems are not straightforward is not a lack of technology. We will be looking at some of the more important elements later on. The novelty of distributed applications, the way in which they defy common perception of computer systems and the sheer variety of technology from which they are built all contribute to an immature disciple, but on that is growing fast.

1.6 SUMMARY

In this opening chapter, we have taken a whistle-stop tour of some of the key concepts and influences in the modern telecommunications world. It is very clear we are looking here at an industry driven by a set of technologies that have come together to enable a hugely powerful set of communications capabilities.

This power is provided by telecommunications providers upgrading their established voice-orientated network to create an information net, and by computer suppliers adding communications to their wares. In doing so, a vast array of technology has to be brought together, integrated and deployed.

Some background to the technology base in our new world of communications is outlined in this chapter, along with the underlying principles in each case. We have examined the fundamental operations of the telephone network, the Internet and distributed computing systems. In later chapters, we will go into more depth. For now, though, you should have a firm grasp on the whys and wherefores of each.

REFERENCES

BCS Trends in IT series (1989) *The Future Impact of Information Technology*. BCS.

Freeman, Roger (1999) *Fundamentals of Telecommunications*. John Wiley & Sons.

Monk, Peter (1989) *Technological Changes in the Information Economy*. Pinter.

Muller, Nathan (1999) *Desktop Encyclopedia of the Internet*. Artech House.

Norris, Mark and Pretty, Steve (1999) *Designing the Total Area Network*. John Wiley & Sons.

Norris, Mark, West, Steve and Gaughan, Kevin (2000) *eBusiness Essentials*. John Wiley & Sons.

Norris, Mark and Winton, Neil (1996) *Energize the Network*. Addison Wesley.

2

Technology Trends

No institution can possibly survive if it needs geniuses or supermen to manage it. It must be organized in such a way as to be able to get along under a leadership composed of average human beings

Peter Drucker

One of the best ways to understand the vast array of technology on offer is to have a framework to put it into. That is what we try to build in this chapter. There are a number of clear and consistent trends that pervade the communications industry. Some are driven by user demand, some by the capabilities of the technology or by supplier innovation and some by the will of government and other national bodies to build a better communications infrastructure

Each of the trends presented here has been evident for some time. So these are not ephemeral moves but rather long-term directions that are likely to persist longer than the specific technologies that enable them.

2.1   FROM NARROWBAND TO BROADBAND

In this first section we look at pressures that are being applied by the users of telecommunications systems for more capacity. Having established why there is a general demand for increasing bandwidth, we move on to look at the principal elements of the network and explain how this can be provided. In this context, we also look at the way in which signals are compressed to provide greater apparent capacity using fixed bandwidth.

Why users need bandwidth

The standard telephone line has been optimised for voice transmission. Even for voice, it does not do a particularly impressive job, as the audio bandwidth is limited to the range 300 Hz to 3.4 kHz. That said, the network fulfils the user's expectation for speech—the demand is for availability and economy, not higher performance.

Despite the dominance of voice traffic, the telephone network has, for some time, been used to carry significant amounts of traffic for which it was never designed. Typically, this occurs when personal computers equipped with modems are used for fax transmission, Internet access and the like. With the telephone network such a ubiquitous and convenient resource, it is only natural for its users to want to exploit it. And so, there is an inevitable demand for faster access and hence for greater bandwidth.

Current modems allow a user to transmit and receive data over the Public Switched Telephone Network (PSTN) at speeds of (typically) 28.8 kbps. This is getting close to the limit for using a modem over the PSTN (about 60 kbps) and, even with compression techniques, different networks are needed if significantly more bandwidth is required (above about 115 kbps). Table 2.1 illustrates the amount of capacity required by a range of telecommunications services.

Table 2.1

Service	Bandwidth Required
Voicemail	Up to 10 kbps
Standard Telephony	Up to 64 kbps
Video Conferencing	128-384 kbps
VCR quality movies	Up to 8 Mbps
High definition TV	16-64 Mbps

As well as making some services possible, higher bandwidths make easy some transactions that would be tedious using the PSTN. For instance, the contents of a single floppy disk would take around 10 min to transfer using a modem connected to the established voice network. This would be reduced to only a couple of minutes on the Integrated Services Digital Network or ISDN. The elements of this network are now explained.

Integrated Services Digital Network

The ISDN is an all digital network that allows a number of services to be carried together on the same circuits. It can be considered to be an extension of the public switched telephone network (PSTN), the key difference being that the analogue local loop of the PSTN is upgraded to give an end-to-end digital connection. This means that the ISDN can readily carry any form of data, such as voice, video and computer files without the need for any sort of analogue to digital conversion.

The initial motivation behind ISDN was to replace the analogue telephone network with a less noisy, digital one. It was, therefore, designed around the same notion as already existed in the PSTN with two separate channels operating at 64 kbits per second. This number springs from the fact that basic, analogue voice transmission requires 8 k samples per second, each of which is encoded as 8 bits.

In the UK and Europe, ISDN is offered in two forms, ISDN2 and ISDN30, where the number suffixes denotes the number of 64 kbps channels that are provided.

ISDN2, also known as Basic Rate Access, provides two 64 K (B or bearer) channels and a single 16 k signalling (D or delta) channel. ISDN30 is also called Primary Rate Access and provides 30 B channels along with a D channel. In the USA, Primary Rate Access is based around 24 B channels, with one D channel. In both cases, Basic Rate is intended for home use, and Primary Rate is meant for businesses.

In practice, there are many instances where some number of channels between two and 30 is required to support a particular service. High quality videoconferencing, for example, requires around 6 B channels. Getting the right speed to suit a wide variety of services involves a technique known as inverse multiplexing

In order to access the Integrated Services Digital Network you need a connection to a digital exchange (such as System X, a Nortel DMS, or Ericsson AXE 10) and for the line length to be within the ISDN transmission limit—approximately 45 db loss over the local link in the UK. In most European countries virtually all exchanges are digital and over 80% of customers can be successfully connected to the ISDN.

The ISDN equivalent of the telephone socket is called the Network Termination Unit or NT1. This is a box that has copper wires going back to the main telephone network on one side and a socket, like the standard phone socket only a bit wider, on the other. ISDN compatible equipment can plug directly into an NT1. If equipment is not ISDN compatible, a terminal adapter or TA, needs to be inserted before the NT1. The TA is used to connect ISDN channels to the interfaces on most current computing and communications equipment (typically RS-232 or V.35).

ISDN standards use a set of 'reference points' (such as the S/T interface between the NT1 and the TA) as a basis for interworking between devices and to define the boundary between the phone network and your private installation (Figure 2.1).

Up to eight devices can be connected to one ISDN line and these can be placed anywhere on a 'bus' connected to the S/T point. There are limits on how far this bus stretches (about 200 m), and power for all devices must be provided.

ISDN adapter cards that plug into a computer, just like a modem, are

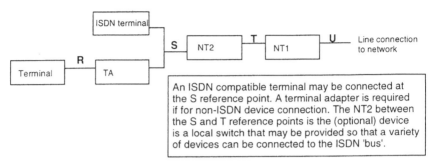

Figure 2.1   The ISDN reference points

readily available. More sophisticated devices, such as those that allow the construction of home networks at the S/T point, are not commercially available at present.

ISDN is offered by local telephone companies, most readily in Australia, France, Japan and Singapore with deployment in Germany very advanced. The whole of Europe is moving towards a common standard called Euro-ISDN. This is a European Telecommunications Standards Institute (ETSI) initiative aimed at allowing full interworking between all European countries.

2.2   FROM VOICE TO MULTISERVICE NETWORKS

With significant growth predicted in the data market, many telecommunications providers have sought ways of extending their mainstream voice services. Initially, they used the underlying voice network as a carrier for data services—the modem connected to a phone line is a prime example of this.

More recently, the idea of a multiservice network, one that can carry voice, video, text and data, has come into vogue. Indeed, one of the main motives behind the Integrated Services Digital Network explained above has been to build a network that can carry all of these with equal ease.

A slightly dismissive evaluation of ISDN would be to say that it gives domestic Internet users a faster, more flexible means of connection along with a spare phone line. Hence it is a multiservice network that satisfies one part of the demand for data services. For higher capacities, there are a number of technologies and services that can be used. Those explained below are widely available and provide the platform for high-speed data services.

Frame Relay
Frame Relay is a product of the digital age in that it exploits the much lower error rates and higher transmission speeds of modern digital

systems. Frame Relay has its roots in the Integrated Services Digital Network and shares much of the ISDN philosophy.

As a development of the PSTN, the ISDN is intrinsically circuit-switched but packet-switching is more relevant for data networking. The first generation of packet-switches (which were based a protocol known as X.25) do not fit the ISDN model of keeping user information and signalling separate and use heavyweight error correction protocols in the comparatively error-free digital environment. Something else was needed in the ISDN to support data services effectively and Frame Relay was defined to fill the gap.

Frame Relay is a simple virtual circuit packet service developed as an ISDN bearer service for use at data rates up to 2 Mbps. It provides both Switched Virtual Calls (SVCs) and Permanent Virtual Circuits (PVCs). It follows the ISDN principle of keeping user data and signalling separate.

A Frame Relay SVC would be set up in exactly the same way as an ordinary circuit-mode connection using ISDN common-channel signalling protocols. The difference is that in the data transfer (or conversation) phase the user's information is switched through simple packet switches—known as frame relays—rather than circuit-mode cross-points.

PVCs can be set up, on subscription, by the network operator. Because the links are fixed, user signalling is neither needed nor provided.

In Frame Relay information is transferred in variable length frames. In addition to the user's information there is a header and trailer. The header contains a 10-bit label agreed between the terminal and the network at call set-up time (or at subscription time if a PVC) which uniquely identifies the virtual call. The format of a one of these frames is shown in Figure 2.2.

Terminals can therefore support many simultaneous virtual calls to different destinations, or even a mixture of SVCs and PVCs, using the

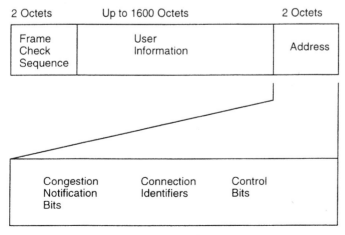

Figure 2.2 Format of a Frame Relay frame

identification facility to identify which virtual call each frame belongs to. Values from 16 to 991 are available to identify the user's SVCs and PVCs (others are reserved for specific purposes—0 is used to carry call-control signalling, 992 to 1007 carry management information.

One of the great merits of the simple data transfer protocol is that it provides a high degree of transparency to the higher layer protocols that are carried. This contrasts with the older X.25 protocol, where the scope for interaction with higher layer protocols often causes problems and can seriously impair performance and throughput.

One issue with Frame Relay is the absence of flow control which leaves the network open to congestion. Congestion ultimately means throwing frames away. Throwing frames away causes higher layer protocols to re-transmit lost frames which further feeds the congestion leading to the possible collapse of the network. Congestion management is therefore an important issue for the network designer and network operator if these serious congestion effects are to be controlled and, preferably, avoided (Figure 2.3).

Given the flexibility of Frame Relay, it is important for the users and the network operator to agree on the nature and quality of the service to be provided. This gives the service provider an estimate of the traffic to be expected, essential for properly dimensioning the network, and it gives users defined levels of service which they can select to match their requirements best. The Frame Relay standards specify a number of parameters which characterise service quality, some relating to the demand the user will place on the network, others specifying the performance targets the network operator is expected to meet.

One of the dominant trends of the last decade has been the rise and rise of the personal computer. From its early use as a word processor it has evolved to become an indispensable part of a company's information infrastructure, and is now almost as common in the office as the telephone. One of the most remarkable features of this evolution has been the

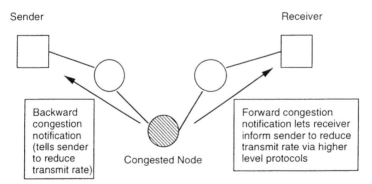

Figure 2.3 Indicating Congestion

associated almost explosive growth in Local Area Networks (LANs) used to interconnect them.

By their very nature LANs are capable of only limited geographical coverage. Companies with LANs in different locations therefore need to interwork them, often over long distances and sometimes internationally, and a confusion of bridges, routers, brouters, hubs, gateways and other devices has been developed to adapt the profusion of proprietary LAN protocols to the available wide area channels used to interconnect them.

Until Frame Relay came along the choice for wide area LAN-interconnection lay between leased lines and X.25. The former tend to be expensive, especially for international interconnection, and not well matched to the bursty nature of LAN traffic. The latter is a complex protocol which tends to interfere destructively with any higher layer protocols being carried, usually degrading throughput seriously, often severely, and occasionally fatally.

Frame Relay's high speed and transparency to higher layer protocols make it an almost ideal choice for interconnecting LANs over wide areas. Hence Frame Relay provides an ideal service for many Internet users and one that is well matched to the type of traffic being carried in small to medium businesses.

SMDS

SMDS, which stands for Switched Multi-megabit Data Service, is a public, high-speed packet data service. Like Frame Relay, it is aimed at wide area to LAN interconnection. The most commonly cited difference between the two is that SMDS is considerably faster—34 Mbps is readily available.

In fact the two are fundamentally different in the way that they operate. This may or may not be of significance to the end-user but it is worth spending a little time on the principles of SMDS.

The first thing to appreciate is that SMDS is a connectionless service. This means that every packet contains full address information to identify both the sender and the intended recipient(s). The SMDS packet is shown in Figure 2.4. SMDS also enforces a point of entry policy such that the network access path can be dedicated to a single customer. Together,

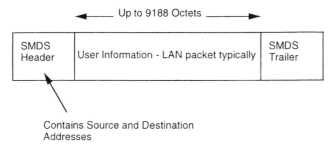

Figure 2.4   The SMDS packet

these mean that addresses can be screened and communication restricted to selected users. So, even though it is based on a public network, SMDS can offer the security of a private network.

Four key service characteristics are defined within SMDS. These are a connectionless service with multi-cast capability, global addressing, a facility for creating virtual private networks and a mechanism that protects users from paying for unused capacity. In terms of access speed, the standard classes used are at 4, 10, 16, 25 and 34 Mbps. Higher rates of 140 and 155 Mbps are planned for the future. With the current range, 10 Mbps Ethernet LANs or 16 Mbps token ring LANs can readily be linked.

A typical SMDS implementation would have a router accepting IP packets which would be attached to an SMDS packet and sent over the SMDS network where the reverse process would take place. The way in which the IP packet (or any other packet, for that matter) is connected to the SMDS packet is called the SMDS interface protocol or SIP. A variety of technologies can operate over SMDS, thanks to the fact that a clear interface is defined. A data exchange interface (DXI) and a relay interface enable a wide range of end devices to connect via SMDS.

One of the best known examples of SMDS is its application in Super-JANET, the high speed academic network in the UK. It provides the backbone that connects around 70 universities and is used for a wide variety of applications, including the transmission of medical images, videoconferencing and education material.

Since SMDS and Frame Relay fulfil much the same function (interconnection of LANs), the choice between them will usually be in terms of price and performance and this is specific to user need and operator tariffs.

ATM

Asynchronous Transfer Mode, or ATM, is seen by many in the telecommunications industry as the basis for a true multiservice network. Unlike SMDS and Frame Relay, ATM is a technology rather than a service. It provides a high speed, connection-orientated mechanism for carrying both, so is not an alternative to SMDS or Frame Relay but is a support to either.

The basic operation of ATM is to route short packets of uniform length (called cells) at very high rates. Each cell is 53 octets (eight bits) long—up to 48 octets of data and five octets of addressing and control information. (Figure 2.5) This uniformity of structure allows the switching of cells to be carried out by static hardware rather than software and it is this feature that underpins the high operational speed of ATM. In action, ATM works by waiting until a cell payload of user information is ready and then adding a cell header before passing the complete cell on to a local switch, where the cell is routed through the network to its destination.

No regard is paid to the content of the cell in this mechanistic part of the

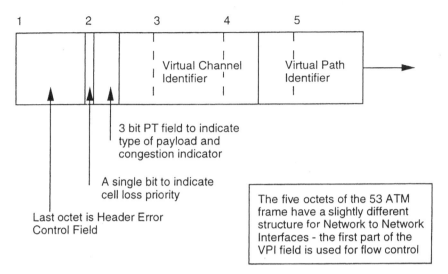

Figure 2.5   ATM

process: a uniform switching is presented to all types of traffic, hence the suitability of ATM to multiservice networks. A variety of control bits are included in the header to secure effective delivery.

The fact that ATM can carry a whole range of different types of traffic—voice, video, text—cannot be ignored. Because each of these types of data has different requirements in terms of delay or error tolerance, there are a number of options defined within ATM for putting the raw information into the cells.

Within the descriptive model of ATM, there is a layer known as the adaptation layer that copes with this. The ATM Adaptation Layer (AAL) sits between any service specific functions and the basic cell assembly layer.

There are four distinct AALs, each defined to support a different class of service (e.g. connection-orientated with constant bit rate, connectionless with variable bit rate) (Figure 2.6). The features provided within each AAL come at a cost of reduced payload—the more sophisticated the facilities required by a service, the more payload is used to provide it.

Over and above these coding options, there are also defined classes of service that allow paths and circuits to be managed effectively.

It was mentioned above that ATM is connection-orientated. Connections are made by creating suitable entries in lookup tables in every switch en route. There are two options for doing this. If the entries are made at subscription time, a Permanent Virtual Circuit (PVC) is created. If, on the other hand, the entries are made at call set-up time, a Switched Virtual Circuit (SVC) is created. The latter place greater demands on switch design and has been a less common option.

AAL category	Supports Traffic type	Known As
1	Constant bit-rate, connection-orientated Timing link between ends	CBR
2	Variable bit-rate, connection-orientated Timing link between ends	VBR
3/4 and 5	Variable bit-rate connection-orientated or connectionless Asynchronous link	UBR or ABR

	Bandwidth Guarantee	Delay Variation Guarantee	Throughput Guarantee	Congestion Feedback
CBR Constant bit rate - well suited to voice and video traffic	✓	✓	✓	✗
VBR Variable bit rate - ideal for bursty traffic such as transaction processing or LAN interconnect, as long as rates do not exceed specified average	✓	✓	✓	✗
UBR Unspecified bit rate - uses any spare bandwidth but gives no guarantees. Useful for standard computer communications	✗	✗	✗	✗
ABR Available bit rate - uses spare bandwidth and gives notification of congestion. Good for email and file transfer	✓	✗	✓	✓

Figure 2.6   The ATM AALs

There is a lot of flexibility built into the way that ATM connections are established. The connections established between two sites, for instance, can be further divided into a number of virtual paths. This allows flexible interconnection of user sites, for instance, a connection may support a link between private exchanges, a videoconferencing link and a Frame Relay service. ATM cards that enable a PC to connect via a local area network have been available since the mid 1990s and ATM network infrastructure is developing quickly, both in the UK and in other countries.

Incidentally, ATM (Asynchronous Transfer Mode) is ideally suited to the support of ATM (Automatic Teller Machines). And this observation is no joke—it illustrates the heavy overload of technical terms in the communications arena.

2.3   FROM DESKTOP TO CORE NETWORK

Everyone is familiar with the telephone: it has been a part of the home environment for so long that its function and capability are known to all. What lies behind the phone, inside the network, has also been described.

The communication elements that make up modern office (both home and business) environments are probably less well known to most people. This section traces the route from the familiar personal computer into one of the data networks described earlier. As well as looking at the various pieces of technology along the way, we consider some of the (growing) expectations of the end user.

Local Area Networks
One of the most dramatic events in computer networking has been the introduction and rapid growth of the Local Area Network (LAN). As the name suggests, this is a means of connecting a number of computing elements together: a personal computer, a printer and any other services that are to be shared between a number of people.

At the simplest level, a LAN provides no more than a shared medium (e.g. a coaxial cable to which all computers, printers etc. are connected) along with a set of rules that govern the access to that medium.

The most widely used LAN, Ethernet, uses a mechanism called Call Sense Multiple Access–Collision Detect (CSMS–CD). This means that every connected device can only use the cable once it has established that no other device is using it. If there is contention, the device looking for a connection backs off and tries again later. The Ethernet transfers data at 10 Mbps, fast enough to make the distance between devices insignificant. They appear to be connected directly to their destination.

Ethernet is one examples of a Local Area Network. There are many different layouts - bus, star, ring—and a number of different access protocols (Figure 2.7).

Despite this variety, all LANs share the feature that they are limited in range (typically they cover one building) and are fast enough to make the connecting network invisible to the devices that use it.

In addition to providing shared access, modern LANs can also give users a wide range of sophisticated facilities. Management software packages are available to control the way in which devices are configured on the LAN, how users are administered and how network resources are controlled.

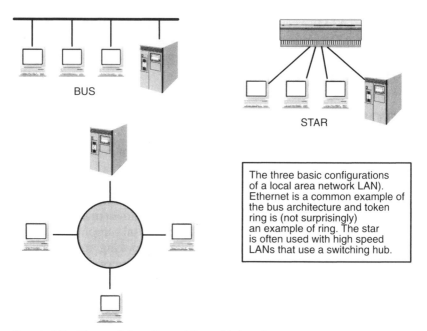

The three basic configurations
of a local area network LAN).
Ethernet is a common example of
the bus architecture and token
ring is (not surprisingly)
an example of ring. The star
is often used with high speed
LANs that use a switching hub.

Figure 2.7   The variety of Local Area Networks

A widely adopted structure on local networks is to have a number of servers that are available to a (usually much greater) number of clients. The former, usually powerful computers, provide services such as print control, file sharing, mail etc. to the latter, usually personal computers.

Routers and bridges
The facilities on most LANs are very powerful. Most organisations do not wish to have small isolated islands of computing facilities. They usually want to extent facilities over a wider area so that groups of people can work without having to be relocated. Routers and bridges are specialised devices that allow two or more LANs to be connected. The bridge is the more basic device and can only connect LANs of the same type. The router is a more intelligent component that can interconnect many different types of computer network (Figure 2.8).

Many large companies have corporate data networks that are founded on a collection of LANs and routers. From the user's point of view, this arrangement provides them with a physically diverse network that looks like one coherent resource—a virtual private network.

Peripheral intelligence
Having a communication path between A and B is only part of the story. Typically, a LAN-connected user will have a network operating system (such as Novell Netware) that allows them to see and connect to any of the

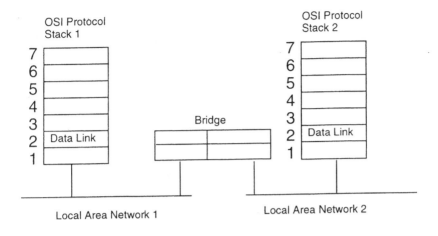

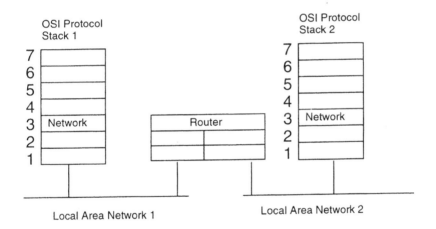

The bridge examines the destination address of each packet
It connect to the LAN at the Media Access Control sub-layer of
the Data Link layer and delivers packets using the Logical Link
sub-layer of layer 2. The router operates at layer 3, and so looks
after the end to end connection, rather than just packet forwarding

Figure 2.8   Routers and Bridges at work

resources on the network. The concept will be a familiar one to many
internet users. The desired resource is selected and, once user name and
password are verified, access to the resource is granted.

The level of control that the end user has in this sort of environment is
somewhat higher than in a traditional telecommunications environment.
It is quite feasible to connect to a number of servers, transfer files between

them, run remote applications, test network elements and download information. The user is very much in control, the servers serve and the network simply delivers.

2.4 FROM WIRES TO WIRELESS

As well as enhanced bandwidth, users have also been demanding greater portability. The needs of an increasingly roving workforce are driving all aspects of mobile communication technology, from simple paging through to full blown remote access to desktop service.

In order to provide mobile services, wireless technology has to be exploited. This means that the limited (and shared) electromagnetic spectrum has to be used, and as efficiently as possible. The breadth of the spectrum is illustrated below.

100 MHz	1G Hz	10 GHz	40 GHz
FM Radio	Cellular Radio PCS	Satellite	Wireless Cable
Television		Microwave Trunks	

The most familiar application on the spectrum is broadcasting—the established television and radio channels. The area of interests from a telecommunications point of view is the 100 MHz to 40 GHz region. Within this, the main allocations are for frequency modulated radio around 100 MHz, for terrestrial television from about 400 MHz to 800 MHz and satellite broadcasting around 12 GHz.

Other major uses of the available spectrum are fixed microwave links between 2 GHz and 12 GHz, satellite links between 4 GHz and 6 GHz and an allocation for wireless cable at 40 GHz.

Within these allocated portions of the spectrum, transmission is moving from analogue to digital format. This allows up to five times as many channels to be carried in a given frequency allocation. Hence there is significant opportunity to offer more services out of a finite bandwidth.

Mobile systems
The mass market service now familiar to many people is the mobile phone service based on cellular radio. The current generation of cellular radio systems work to the European Global System for Mobile Communications (GSM) standard which uses the 900 MHz frequency range and Personal Communications Systems (PCS) using GSM in the 1800 MHz frequency range. The GSM standard has been widely adopted throughout Europe. In Japan and the USA there are comparable (but not compatible) standards (Figure 2.9).

A GSM service works in much the same way as the established analogue

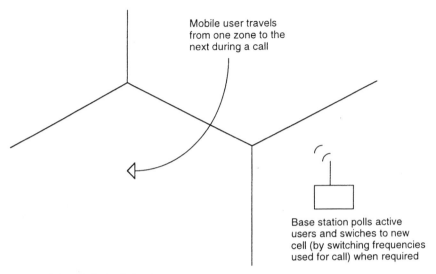

Mobile user travels
from one zone to the
next during a call

Base station polls active
users and swiches to new
cell (by switching frequencies
used for call) when required

Figure 2.9   Cellular telephony

mobile system. First, the handset makes contact with the nearest base station, which then feeds into a base station controller. This establishes where the handset is currently located. The base station controller is linked to a mobile switch centre that can interface to other mobile switch centres as well as to the fixed network. As a call progresses, a path to the handset is maintained through whichever base station is appropriate. Call continuity is effected through a visitor location register which ensures that all active calls are tracked and connected to the right base station. The base stations themselves are deployed in a cell formation over the geography so that the handset is always contactable (Figure 2.10).

Over and above the tracking aspects of the system, there are a number of other controls. An equipment identity register is kept so that it is known what can and cannot be connected to the network. An authentication register contains details of acceptable subscriber identification modules (these are the cards inserted into the mobile handset). A home location register contains information on the class of service that a particular customer is allowed.

Digital technology brings a number of advantages when applied to mobile communications:

- Bandwidth can be used more effectively. A digital signal can be coded at a low bit rate (e.g. 13 kbps for GSM) and modulation techniques can ensure that minimal frequency space is used.
- Handsets can be low power. The robustness of a digital signal (a carrier to interference ratio of 10 dB is quite acceptable) means that low power levels can be used.

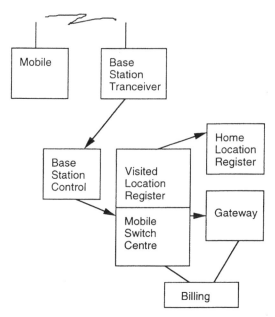

Figure 2.10 Structure of a GSM network

- Security of information. A digital signal is easier to encode and more difficult to intercept than an analogue one. Techniques such as frequency hopping and key encryption can be used.

- Data is easier to accommodate. Just as with the ISDN, there is no need to swap between analogue and digital formats.

- Extra service. For instance, the short message service (SMS) is incorporated into the GSM standard. This adds radiopaging facilities into the phone capability.

The GSM standard

The GSM approach is to use a bandwidth of 200 kHz for each separate channel and, within each of the channels, to accommodate a bit rate of 270 kbps. The bit stream is divided up into a series of frames with the pattern repeating at a rate of 217 times per second. Each of the frames has eight slots in it. Each of the slots has a duration of 0.577 ms and, at the given bit rate, this corresponds to 156.25 bits.

The bits that are contained within each slot are made up of some overhead bits, some fixed patterns and, predominantly, the bits allocated for the signal itself. The fixed patterns are there to allow the equalisers within the overall system to adapt to the characteristics of the radio signal and the overhead bits allow control over the handset to base station dialogue to be effected.

GSM encodes speech signals at a rate of 13 kbps and this is protected against error by adding some extra bits into the transmitted stream. The resultant gross data rate is 22.8 kbps.

The frequency allocations for GSM are 900 MHz—960 MHz for base stations and 890 MHz—915MHz for the handsets. Also reserved are 1800 MHz to 1880 MHz for base stations and 1710 MHz to 1785 MHz for handsets.

The existing implementation of GSM is now widely deployed across Europe. The next generation of digital mobile systems, known as Universal Mobile Telecommunications System, is in the process of being defined. The basic objective is to start introducing UMTS, which would allow communication at rates up to 2 Mbps from the year 2000.

Paging systems

A very basic communication mechanism is offered through paging. In effect, the pager allows a one-way message, from caller to called party. The current generation of pagers have sophisticated messaging and alert facilities but the basic service is the same.

Pagers are very popular, primarily due to the fact that they tend to be inexpensive. In Singapore, some 12% of the population regularly carry a pager. Forecasts indicate an installed base of 100–120 million pagers by the end of the decade (compared with around 55 million in 1997)

The European standard for digital paging is called Ermes. This is intended to provide a Europe wide messaging system. In the USA, Skytel operate a wide area paging system.

One of the attractive features of paging systems is that they can operate effectively using spare capacity on broadcast system transmitters. Many broadcast transmissions employ a subcarrier which can be transmitted easily and which does not interfere with the signal being broadcast. Pagers require very low data content per message. Hence, the capacity of these channels is sufficient to support a large number of pagers.

Satellite systems

These are extensively used for broadcasting and also have a part in the public network, although this is considerably less than it once was.

Nearly all of the satellites that are used in telecommunications are in geostationary orbit. This means that they are fixed in position, relative to the Earth and can be accessed with fixed antennas that are pointed at them. The geostationary orbit places the satellites at a height of 35, 786 km above the Earth. There is an international agreement that satellites should be place in slots that confine them to 2 degrees of longitude and a specific range of frequencies used. Such close spacing means that sharp antenna beamwidths are needed if interference is to be avoided.

The role of the satellite is that of a large relay station with limited on-board switching capability. A satellite will have a number of transponders, which take in a fixed frequency range, usually around 36 MHz or 72 MHz wide, for retransmission.

Satellites were once widely used for trunk transmission across the oceans. The installation of a series of undersea optic fibres has superseded them in this function: the fibres allowed the 600 ms introduced with satellites to be removed.

Satellites have developed a niche in business communications and in mobile services with Very Small Aperture Terminals (VSATs). These can work with leased capacity on satellites to give businesses a flexible corporate communications network that is well suited to providing service in remote areas. VSATs are widely used for applications such as credit card authorisation and ordering applications.

VSATs are typically operated by a contractor who will rent a frequency allocation (or even a complete transponder) from the satellite provider. Within the rented frequency band, they will then allocate a time slot for each connected terminal.

The VSATs themselves operate with a small antenna. A network of terminals can be built to operate off a hub that acts as a distribution point for information (in the form of packets) between the terminals in the network.

One of the developments in this area is Demand Assigned Multiple Access (DAMA). This allows extra capacity to be provided to a particular terminal, as needs be. So, if a number of videoconferencing calls are being carried on the network, extra bandwidth will be assigned to the terminals carrying the load.

Mobile satellite systems

Satellite systems are well suited to providing mobile services. Not only do they give broad coverage, they are also often the only alternative in isolated or hazardous areas.

The long established service organisation for mobile satellite systems is Inmarsat. This is a consortium of 74 nations providing service via 25 000 terminals, mostly to ships.

More recent developments are proposed: Low and Medium earth orbit systems (LEO and MEO, respectively). These include the Iridium project (led by Motorola with the aim of providing a global communications service) and the Odessey programme led by TRW.

As LEO and MEO are not geostationary, the satellites move relative to the Earth. This means that it is possible to get better global cover and (due to the lower orbit) to reduce the round trip delays inherent with geo-stationary satellites. In turn, it is possible for small handheld terminals to be used instead of dish antennas.

The idea behind the Iridium project is that each of the 60 satellites in orbit would project around 50 beams onto the Earth, thereby giving around 3000 cells. As the satellites move relative to the earth so the cell patterns on the ground move, so part of the system design is a handover mechanism. This is clearly well suited for providing a mobile communication network. Iridium would provide hundreds of thousands of channels with transmission in the 161 MHz to 1626 MHz region.

2.5   FROM SWITCHED TO INTELLIGENT NETWORKS

Until the 1970s all Public Switched Telephone Networks (PSTN) were based on electromechanical switches, mainly step-by-step Strowger and crossbar systems. The main characteristics of these switching systems was that signalling was carried over the same transmission path as the analogue voice signal, and the control of the switching systems was highly distributed, being integrated with the switching fabric itself.

This meant that the systems were intrinsically reliable in that component failures, though common, had a limited effect on service, typically affecting only one call. But it also meant that the nature of the service offered to the user was closely bound into the switches and could not be altered or enhanced without complete redesign.

However, starting in the late 1970s, the development of computer technology led to dramatic changes in the PSTN. As analogue transmission was progressively being replaced by digital transmission the conversion between analogue and digital moved so that it was being performed at the serving local switch. The electromechanical switches were replaced by digital switching systems with computer control (Stored Program Control, SPC) using powerful message-based common-channel signalling for conveying control information between the processors. In effect this creates a network consisting of two logically distinct subnets, a switched voice subnet and a signalling subnet.

This network became known as the Integrated Digital Network, or IDN. The software control of services together with the powerful signalling promised great flexibility in managing the network and creating and managing new services. However, this promise was only realised to a limited extent because it quickly became clear that making software changes was in practice far from trivial. Nonetheless, the adoption of standard mechanisms for call control, most notably CCITT signalling system No. 7 (C7), provided a basis for greater network flexibility.

With the IDN, a call using C7 would progress through the network as follows:

- The caller goes off-hook, gets dial tone and dials the destination number. When their Local Exchange (LE) has received enough digits to be able to route the call (probably the area code) it sends an Initial Address Message (IAM) to a Trunk Exchange (TE) selected by the routing decision. This message contains the dialled digits and any special requirements for the call path (such as that an end–end digital path is needed and routes containing speech processing equipment must be avoided). The near end TE then makes its own routing decision and sends a similar IAM to a terminating (or intermediary) TE.

- This TE decides that it needs to receive more digits before it can complete the call (this would be the balance of the national telephone number, that is the local part following the area code) and returns a Send N Digits message to the originating TE which relays it to the calling LE. N is the number of digits needed to complete the call.

- When all the digits have been received from the caller the local LE sends a Subsequent Address Message (SAM) to the originating TE and thence to TE at the far end. This message contains the required additional dialled digits. The remote TE then sends an Initial and Final Address Message (IFAM) to the called party's LE. The IFAM contains all the dialled digits. On receipt of the IFAM the destination LE checks the state of the called line and if FREE sends an Address Complete Message (ACM) back to LE A via the trunk exchanges. At this point the call path is connected, ring tone applied to the caller and ringing sent to the called party.

- When the called party answers, the terminating LE sends an Answer (charge) message back to the originating LE via the trunk exchanges. The originating LE begins charging for the call and conversation takes place, and at some later time the call would be cleared. The exact sequence of signalling messages used to achieve this depends on whether the calling or called party clears first.

A similar sequence would hold for an Intelligent Network call. The main difference would be that signals could be intercepted before reaching the local exchange and referred to a decision point so that appropriate actions could be taken.

The actual network intelligence is additional software provided above basic switch control for path set-up and release to make service functionality independent of the basic switch. The level of network intelligence has increased significantly since its inception in the early 1990s. It looks likely to continue progressively to revolutionise the concepts behind service delivery to customers by providing much greater flexibility and service customisation.

The introduction of this capability within the network will progressively reduce the function of the switching elements to the point where probably only the most basic simple calls are handled totally within the switch software. This represents a fundamental reversal of the trends in past years, where switch manufacturers have implemented increasingly complex service logics within the switch software.

The availability of a standardised intelligent network architecture defined by CCITT (now ITU-T), standard service definitions, standardised interfaces between switching elements and processing equipment, and standardised computing platform environments will make it possible to construct a sophisticated intelligent network from hardware and software elements obtained from multiple vendors. The importance of standards is already evident in the adoption of C7 and will grow in this area, as they already have in others (both as an enabler for the user and as an essential feature for the vendor).

Elements of the intelligent network

The major building blocks of the IN architecture are the Service Switching Point (SSP) and the Service Control Point (SCP). The SSP is a digital switching system whose call control software has been restructured to separate the basic call control that looks after fundamental switching operations from the more advanced call control needed for intelligent services. This restructuring involves large-scale software development, a fact that continues to shape the evolution of IN deployment (Figure 2.11).

In particular the basic call control process has been redesigned to incorporate Points In Call (PICs) and Detection Points (DPs) as defined points in the basic call control state machine at which trigger events, such as recognition of the Calling Line Identity, may be detected and call processing temporarily suspended whilst reference is made to the SCP to find out how to handle the call.

The SCP is a general purpose computing platform that implements the advanced call control needed by IN services and stores the information that defines each customer's service. It has to be fast in order to provide the rapid response and high throughput needed, and reliable. In a practical

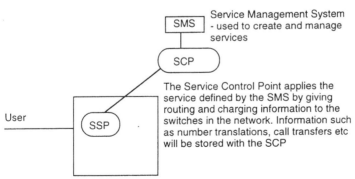

Figure 2.11 Elements of an IN

network more than one SCP would be provided to meet the requirements for throughput and availability. There may indeed be a dozen or more.

Requests for a specific service are transferred from the customer's Service Switching Point (SSP) by the signalling system to the Service Control Point (SCP). This determines the network actions required from the service logic of the required service (e.g. number translation time of day routing). It may also arrange connection of the customer to an intelligent peripheral device (e.g. for digit collection or voice prompted customer menu selection), or collect information from other remote service control points concerning service logics for the call destination. Once the appropriate actions have been determined, the signalling network is used to pass specific instructions for providing the requested service back to the service switching points.

The Service Creation Environment (SCP) provides tools to the network provider to allow rapid creation of new types of services from reusable software components, while the service management system provides operations support, allowing the updating of SCP information relating to particular customers or particular network services. Implementation of such intelligent network architectures will commence from around the middle of the decade, and will be well advanced by the year 2000.

Virtual private networks
A Virtual Private Network or VPN is similar to a Private Network in that it appears to be custom designed to meet the needs of an organisation (Figure 2.12). In reality, it is built from a range of components, some owned, some public and some leased.

The user of the VPN is obliged to supply their requirements in much the same way that they would were they building their own network. The supplier has to deploy their equipment and expertise to satisfy the requirements of as many paying customers as they can.

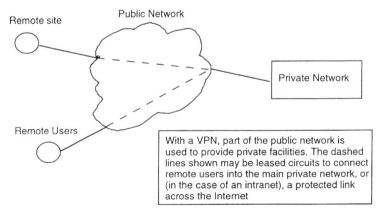

Figure 2.12 A typical VPN

Over the years, VPNs have appeared in a number of guises, Enterprise Networks and Intranets being two of the more popular. The latter, which is a special case of a VPN (one constrained by the technology used on the Internet), is of growing importance.

In reality, the VPN is not a packaged commodity but a reconfiguration of resources to meet the user need. To illustrate, a VPN can be created by a supplier by configuring a set of closed user groups and selling them at a flat rate to customers. Little technical work takes place, yet a network that is customised to the needs of the purchaser is created.

At present, the bulk of the VPN market lies in the USA. VPN services over there account for around $4.1B, compared with only $400M in Europe. This is expected to change, though, and the predicted market size for VPNs is $21B, with 60% of this in Europe.

It is possible to illustrate the user motivation for a Virtual Private Network, by taking a simple example. Suppose an organisation with five sites wants to interconnect its installed base of private exchanges or routers. It can do this quite readily by purchasing leased lines: 10 links are required and each site has four ports.

If the organisation is expanding, a problem quickly arises with the private network solution. As each new site is incorporated in to the network, a connection to all of the existing sites is needed and another port has to be provided at each termination. By the time the network has grown to 75 sites (which is not uncommon), 74 ports are needed at each location to ensure full connectivity and the number of links has shot through the roof! Clearly, the demand on network resources grows faster than its useful expansion, making the approach both expensive and difficult to maintain.

The use of part of a switched network, provided by a third party, makes growth less of a problem as each new site requires only one port—its connection to the VPN. The problem of managing the switched network and ensuring that it has the capacity to connect each of the sites as if they had dedicated links between them is left with the supplier.

Many telecommunication companies offer VPNs. Initially, these were voice offerings and gave the customer facilities such as a private network numbering scheme, fast dialling, secure access and selective routing. More recently both voice and data VPNs have become available, with Intranets a popular example of the latter. Typically, an Intranet vendor would provide a number of customer routers (to which local networks can be connected), a firewall (to restrict access) and a variety of services to cope, for instance with numbering and naming.

2.6 COMPUTER TELEPHONY INTEGRATION

Computer Telephony Integration is a broad term covering the integration of telephony functions with computer applications. It takes many forms in practice, from the combining of a single phone with a personal computer to the addition of computer interface to private exchanges and automatic call distributors.

CTI is probably most widely used for distributing incoming calls within a call centre (a dial-in support and marketing facility staffed by customer agents). In this instance, the calling line identifier provided by the PSTN is intercepted, analysed and used to switch the call to the most appropriate recipient. In addition to this, details of the caller can be retrieved and placed onto a screen prior to any dialogue.

The integration in CTI is very much through interfaces. These provide the flexibility to provide a wide range of capabilities. A computer user can configure the local voice switch to place voice calls where they want and incoming calls can trigger a set of relevant data to be retrieved from a computer and placed with a relevant person.

The key interface standards for CTI include a set for PC based applications (such as TAPI and TSAPI) and a set for switch equipment (such as CSTA and SCAI).

TAPI, or Telephone Application Program Interface, is an application programming interface devised by Intel and Microsoft that allows independent developers to provide applications for CTI. Similar in function (and probably more widely used in practice) is TSAPI, which was devised by AT&T and Novell.

The interface standards for switch equipment are also similar. Switch/Computer Applications Interface, like CSTA, are standards that allow developers to produce CTI applications that reside on a network connected switch. The difference between CSTA and SCAI is that the latter has been developed through standards bodies and is intended for both public and private network operators. Both SCAI and CTSA are aimed at network owners operators, rather than end users.

Despite its rather grandiose name, CTI is, at present, something of a specialist area. Nonetheless, it highlights the way in which telecommunications and computing can combine to automate a particular job.

2.7 SUMMARY

In this chapter, we have looked at half a dozen of the major trends in communications. Specifically, we have explained the move from

- From Narrowband to Broadband, driven by greater volumes of communication and hence demand for more network capacity.

- From Voice to Multiservice Networks, driven by a desire to support the growing market for data along with the established one for voice.

- From Desktop to Core Network, where the end user is provide with all of the elements needed to control the resources at their disposal to deliver the desired information.

- From Wires to Wireless, driven by the 'virtual worker' who wants to have access to their network from anywhere.

- From Switched to Intelligent Network, so that new services can be added without having to change the configuration of the whole network.

Finally, we looked briefly at the current state of Computer Telephony Integration as an indicator of the way in which computing and communications technologies are coming together.

REFERENCES

Atkins, J. and Norris, M. (1998) *Total Area Networking.* John Wiley & Sons.

Held, G. (1998) *Internetworking LANs and WANs.* John Wiley & Sons.

Norris, M. (1999) *Understanding Network Technology: Concepts Terms and Trends.* Artech House.

Van Duuren, J, Kastelein, P. and Schoute, F. (1996) *Fixed and Mobile Telecommunications.* Addison Wesley.

Walters, R. (1998) *Computer Telephone Integration.* Artech House.

Whyte, W. (1999) *Networked Futures — trends for communication systems development.* John Wiley & Sons.

3

Key Telecommunications Concepts

One day, every town in America will have a telephone

Mayor of a small American town, circa 1880

The telecommunications network is probably the largest human construction of all time. Every day it carries many millions of calls between virtually every country in the world. Calls very rarely fail and, once set up, the quality of the link is usually perfect. Just about everyone knows how to use the telephone because the complexity of the huge machine it is connected to is masked behind a simple and intuitive interface.

This impressive feat of large-scale engineering was no fluke. Ever since Alexander Graham Bell summonsed Mr Watson, designers have been hard at work making the telephone network operate better, faster and more efficiently.

So, in this chapter we are going to focus on the voice network, concentrate on phone calls and, along the way, explain some of the key concepts that underpin telecommunications.

3.1 BACK TO BASICS

We will start with the absolute basics of any call: how to transfer a signal from A to B. There are three elements that we need for this: how to transport the signal, how to get it to the right place and how to control the call.

Transmission

The very first thing to do is to ensure that a signal can be carried between two parties. In telecommunications parlance, this is referred to as transmission, and is a highly developed specialism in its own right. There are three key aspects to transmission and these are the way in which transmitted information is organised, the way in which it is presented and the medium over which it is sent. We will now look at each of these in turn.

One of the key issues with just about everything to do with telecommunications is scale. Transmission is no exception and engineers have had to devise a multiplexing scheme that can send large amounts of information across a network in an orderly fashion. The early systems, for analogue traffic, were based on Frequency Division Multiplexing (FDM) (Figure 3.1), where multiple signals are distributed in frequency space, just like domestic radio stations. This worked but was not the most efficient way of sending telephony signals down a cable.

Nearly 50 years later, the major step on from FDM was made. This was Pulse Code Modulation (PCM), where a digitised voice signal is sent down a cable in one of a number of pre-allocated time slots. Typically, a time slot of 125 m would accommodate 30 speech channels, each of 64 kbps. Two spare slots would be included to allow for synchronisation and signalling (Figure 3.2).

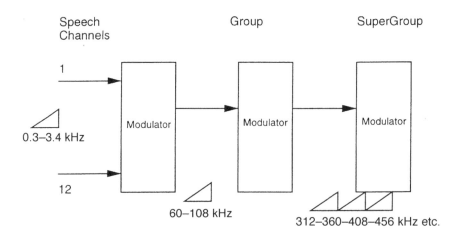

The input speech channels are modulated with carrier signals from 60 to 108 kHz. Hence each one is carried in the frequency spectrum at the carrier frequency.

Figure 3.1 FDM

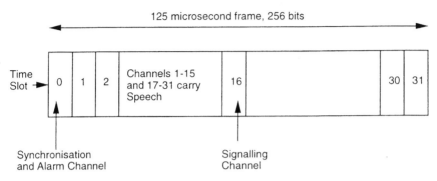

Figure 3.2 PCM structure

As well as removing the problem of crosstalk between signals in the cable, PCM opened the way for a series of enhanced transmission systems, all based on the same principle of digital interpolation.

Since the advent of PCM, a couple of flexible and high capacity transmission systems have been developed and deployed. These are both based on the same basic ideas as PCM—digitised signals carried in time slots—but are organised to carry very high volumes of data. The Plesiochronous Digital Hierarchy (PDH) defines a structure of tributary bit streams that are aggregated into higher order streams. So, for instance, four basic 2048 bps PCM multiplexes would be interleaved to form a 8448 bps or E2 and 16 of these E2s would be used to build a 140 Mbps E4 stream. In the USA, the comparable numbers are four basic 1544 bps form a D2 stream of 6132 bps and so on (Figure 3.3).

The word 'plesiochronous' in PDH means 'almost the same rate' and this indicates that there are slight timing variations when the higher order D2 and E2 frames are assembled. Because of this, it is difficult to access a particular tributary without demultiplexing the whole lot. For this reason, a successor to PDH, known as Synchronous Digital Hierarchy (SDH, or SONET in the USA) was introduced. This has a similar structure to PDH but is more flexible in that it allows access to an individual bit stream within the higher order multiplex. Hence, it is possible to extract a specific signal, and pieces of equipment known as Add-Drop Multiplexers (ADMs) are available to insert/extract specific SDH/SONET streams (Figure 3.4).

As well as devising a structure for the bits going down a wire, transmission is concerned with the way in which those bits are represented. Line signals can be characterised in terms of their power density spectrum and their direct current component. A good line code seeks to minimise undesirable transmission features, such as a high DC component (which wastes power) or few transitions (which make it difficult to synchronise a digital signal.

The three most commonly used coding schemes in communications are

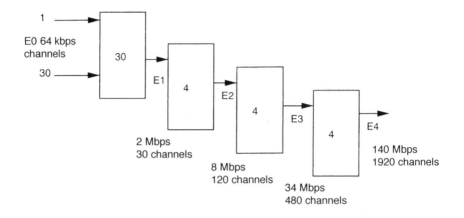

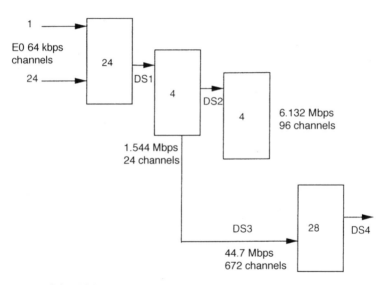

Figure 3.3 PDH

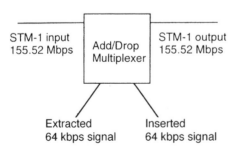

Figure 3.4 SDH/SONET add/drop multiplexer

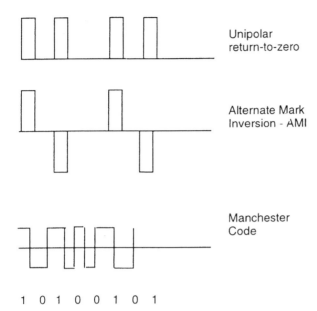

Unipolar
return-to-zero

Alternate Mark
Inversion - AMI

Manchester
Code

1   0   1   0   0   1   0   1

Figure 3.5   Popular line codes

Alternate Mark Inversion (AMI), High Density Bipolar 3 (HDB3) and Manchester encoding. AMI inverts the polarity of every other '1' in a digital stream, hence removing any DC component. HDB3 built on this by ensuring that no more than three sequential zeros are sent, hence easing synchronisation. With Manchester encoding, every digit is coded as a transition from + to—(as opposed to the +1 or –1 coding used with AMI and HDB3). This gives good synchronisation information but generates a wide power spectrum (Figure 3.5).

There are a variety of media with which the coding and aggregation schemes outlined are used. Traditional copper cables are used in local networks (both Local Area Networks and for the delivery of domestic telephone service) but optical fibre is the most widely used medium for bulk transport. Increasingly important is the use of satellite and radio transmission, although the available spectrum puts limits on this.

Switching

Being able to transfer a signal from A to B is all very well but a real telephone network needs some mechanism for sending signals from A to B or C or anywhere else for that matter. And this is where the switch plays its part. Its basic function is that it should be able to connect any of its inlets with any of its outlets. This sounds very simple but it soon becomes apparent that every established path through the switch blocks the way for the next one. And so some thought has to be put into how best to arrange the switch so that it permits as many simultaneous calls as possible. This activity is called trunking, or grading, and it is explained further below.

In order to understand a bit more about what a switch does, it is helpful to follow the sequence of a typical call. The first thing that happens is that the caller lifts the handset. For this action to be detected by the local exchange, we need to trigger a signal (in practice, the completion of a direct current loop) to be associated with the caller's action. Now that the caller's intent is known (and they have been identified by their position on the exchange's incoming links), the number of the called party can be received and dial tone is sent. This is not strictly true, as the exchange has first to allocate resource to deal with the call (such as storage space for the dialled number) but this happens so fast that no delay is perceived between picking up the phone and the exchange being ready.

The dialled digits are analysed by the exchange to see if the call is local, national or special (e.g. non-geographic numbers, freephone services). How the last of these three are treated is explained in some detail in Chapter 7. For the common or garden call to a number out of the telephone directory, the exchange sets up a path using an in-built routing algorithm and (typically) passes this to the next exchange on the way to the call destination.

When the call is answered the speech path between the two parties is established and call set-up is complete. The exchange releases the short-term resources required to establish the call and just maintains a supervisory function so that the speech path is released when the call is cleared.

The earliest exchanges were based on electro mechanical strowger equipment. These dealt with each digit as it was received (typically from a rotary dial as a series of loop disconnect pulses). And so the call had to be set up in a piecemeal fashion, step by step, as each digit was received. Modern exchanges are large, powerful computers and they operate in a completely different way. They receive all of the dialled digits (usually from a keypad as a series of multi-frequency tones) before deciding what to do. They can, therefore, take more sophisticated routing decisions based on the whole number, as we will see Chapter 4

Looking at a bit more detail at one of these modern switches, they are comprised of two distinct components, a time part and a space part. The former takes part of a signal (in the form of bit stream) and copies it to another bit stream, the latter physically diverts the path of a bit stream. The most common switch configuration is a time-space-time (TST) switch. In this arrangement, an input time switch will connect an input time slot to the first available space switch port, while the output time switch connects the chosen time slot from the space switch to the required outgoing time slot. Thus, calls through the whole switch can be routed via any convenient time slot (Figure 3.6).

In order to serve a large number of people efficiently, a lot of thought

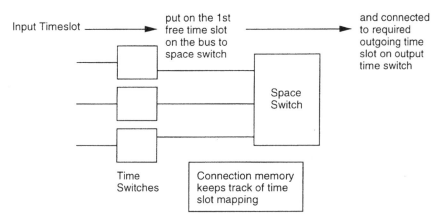

Figure 3.6 Switch Structure

has to go into the design of the switch. It would be infeasible to have one that connected everyone to everyone else. This might be wonderful from the user's point of view as calls would always go through with no delay, but it would cost far too much. A lot of engineering has gone into the telephone network and engineers do not like waste. Techniques for providing good service with a manageable level of equipment have been honed over the years. Some of the keys ideas that have been built up to achieve this are concentration and grade of service.

When a customer's line arrives at a local exchange, there has to be a physical termination. All of the exchange functions (see below) are provided on a line-by-line basis. This does not mean that every incoming line has to be connected to the switch matrix, though. Typically, an exchange will concentrate X incoming lines onto a switch with Y inlets and, on the output side, expand the Y switch outlets to X outgoing lines (Figure 3.7). The ratio of X/Y is arranged to provide reasonable service given typical phone usage rates and patterns.

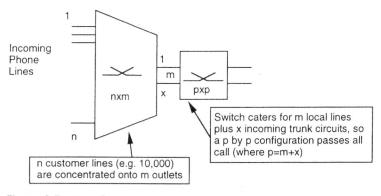

Figure 3.7 Line Concentration

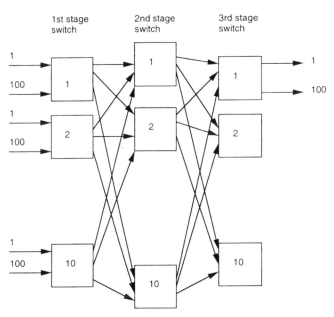

Figure 3.8   Splitting the Switchblock

Further economies can be effected by splitting the switch into a number of interconnected stages. It is easy to illustrate the sense in this. A switch with 1000 inlets and 1000 outlets would have 1 million crosspoints. By splitting it into three interconnected stages, each with ten 100×10 switches, yields a crosspoint count of 10(100×10) + 10(100×10) + 10(100×10), a total of 21 000. This option reduces the number of simultaneous paths through the switch—a condition known as blocking—but as with the concentration option, a balance of economy against performance can be made (Figure 3.8).

If we think just a little more about the problem of serving a huge number of people with a viable amount of equipment, it is clear that the same sort of structure applied within a switch need to be applied to the collection of switches that make up the network. We have already mentioned the 'local exchange' and most people are familiar with the idea that an exchange serves customers in a specific area.

In order to set up national and international calls efficiently, there needs to be a hierarchy of switches, and this is the case for all of the national telephone networks. Local exchanges are fairly isolated from the main network, connected to customers on one side and a trunk or transit exchange on the other. Typically, these transit exchanges will be completely interconnected. Hence, no call need go through more than two local exchanges (the callers and the called party) and two transit exchanges. The function of the transit exchange is to interpret dialled digits and route the call appropriately (Figure 3.9).

The balance and placement of local and transit exchanges depends on

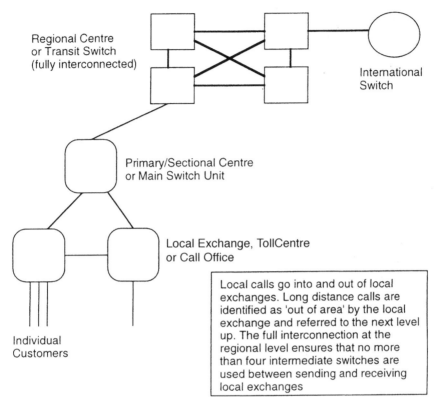

Regional Centre
or Transit Switch
(fully interconnected)

International
Switch

Primary/Sectional Centre
or Main Switch Unit

Local Exchange, TollCentre
or Call Office

Individual
Customers

Local calls go into and out of local
exchanges. Long distance calls are
identified as 'out of area' by the local
exchange and referred to the next level
up. The full interconnection at the
regional level ensures that no more
than four intermediate switches are
used between sending and receiving
local exchanges

Figure 3.9   Switch hierarchy

the geographic distribution of customers and the pattern of calls. Again,
this is an area of specialist study in telecommunications

Before moving on from our introduction to switching, it is worth
mentioning that there are some general functions that exchanges provide.
These basics include power feeding, the provision of dial and ring tones,
protection, signalling etc. and are referred to as BORSCHT—battery feed,
over-voltage protection, ringing, supervision, coding, hybrid and testing.

- Battery feed—a 50 V direct current feed that provides the power source
  used to detect off-hook and to power a simple telephone.

- Over-Voltage protection—a safeguard placed at the periphery of the
  exchange to combat lightning and other electrical problems.

- Ringing—a 75 V, 20 Hz feed interrupted to provide a distinctive
  cadence.

- Supervision—the exchange's monitoring of calls in progress, where
  current state is checked against previous state and appropriate action
  inferred

- Coding—conversion from the analogue signals that some phones use to the digital format (PCM) that all exchanges use.

- Hybrid—a two to four wire conversion device that sits at the periphery of the exchange to all two-way transmission.

- Testing—the ability to check that a particular line is working by proving the transmission path to and from the exchange.

In general, the exchange needs to provide all of these functions for an analogue telephone line. With digital lines, many of the functions are included in the customer's equipment.

Signalling

The term 'signalling' describes the passing of information between switches (these being the exchanges and routers within a network) to set up and control calls. The role of the earliest signalling was simply to gain the attention of an operator who would then connect the call. With the replacement of operators by automatic switching equipment, the address information has to be conveyed electronically, rather than by a human voice and this led to the signalling systems that are used in modern telecommunications.

By and large, we can partition signalling into two varieties. First is the communication between a customer and the network. This tends to be fairly simple as little more than dialled digits plus a few supplementary codes are exchanged. The second type of signalling in telecommunications networks is altogether more complex. This is when control messages are sent across the network (e.g. from one exchange to another). In this instance, a richer set of signals is required, as there are many functions that need to be monitored or controlled.

Most end customers will be familiar with telephony signalling as something you can listen to. The early 'loop disconnect' method (which simply broke the line current to pass digits) had its characteristic clicks and the more recent Multi-Frequency (MF) method (which used two voice-band tones for each button pressed) was frequently mimicked by people who wanted to by-pass the keypad. Both loop disconnect and MF are examples of channel associated signalling—the digits use the same path to the exchange as the call.

In the early days of telephony, MF systems were also used for inter-exchange signalling. This was appropriate for step-by-step exchanges equipped with Strowger switches. With modern stored program control switches, things are a little different. With each exchange having, in effect, a computer to control call processing, the type of signalling deployed across the network can be very different. The channel associated signalling, familiar to the end customer, can be replaced by common channel

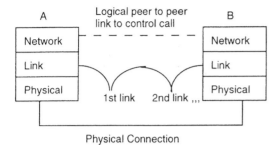

Figure 3.10   CCS in the ISO structure

signalling, a data link between call control processors of interconnected exchanges. And in doing this, the signalling (or control) path can be kept separate from the transmission (or user) path.

Call control information is passed over the inter-exchange data links in the form of messages to establish calls and to request supplementary services. Because the control and user paths are separate, signalling information can be exchanged at any time without interference.

In a digital stream, the signalling information looks much like any other information. It can only be deciphered when there is a predetermined structure. Common channel signalling systems are structured in accordance with the layered architectural model for data transfer recommended by the standards body, ISO (Figure 3.10). In terms of this model, a signalling system is contained in the first three layers. The first provides the physical transmission path between exchanges. The second carries the messages from point to point and protects them from errors and the third is the content of the messages themselves.

The ubiquitous common channel signalling system used by all Telcos is known as Common Channel Signalling Systems No. 7 (or simply, C7). This standard was developed by the CCITT (now ITU-T), who seek to harmonise international telecommunications (see Chapter 9). C7 provides a comprehensive set of inter-exchange messages and is described in some detail in the next chapter.

3.2   SOME PRACTICAL ISSUES

Of course, it is not enough just to have a basic structure that carries calls. A real telecommunications service must ensure that those calls arrive safely, that the network is efficient, throughput has to be assured and so on. There are a whole host of issues that need to be attended to in an industrial-strength network and we look at some of the key ones in this section.

Local Access

Telephony is, in essence, a personalised service—despite the quote with which we opened this chapter! This means that there has to be a connection from the telephone exchange to each and every customer. In practice, this means that many millions of houses, offices and flats have to be connected and, from the telecommunications operator's point of view, this is a major investment that has to be protected.

Typically, local access takes the form of a pair of copper wires. These carry the power feed from the exchange as well as the call and its associated signalling into the network. For many years, the local access link was constrained to a bandwidth of about 3 kHz. This was enough to carry voice and (with a modem), data of up to 56 kbps.

Recent developments have extended the usefulness of the local line. The Integrated Services Digital Network, now widely available, provides a basic rate access of 128 kbps (2 × 64 kbps) over the existing copper pair. Digital Subscriber Line (DSL) technology allows up to 2 Mbps local access, albeit with some restriction imposed by the length of the line. There are several varieties of DSL—asynchronous (where the rate is higher downstream than upstream), Very high bit rate (for wide band services) and Rate adaptive (which adapts to a wide range of loop conditions). In each case, a sophisticated coding scheme (2B1Q) is used. Both ISDN and DSL required extra equipment at each end of the customer to exchange link.

Numbering and Addressing

Two main types of numbering scheme are used in public telecommunications networks—one for public switched telephony and one for public data networks. The schemes for telephone and data networks are defined in two ITU-T standards, E164 and X121 respectively. Although different in structure, these share the common purpose of unambiguously identifying discrete network terminating points (Figure 3.11).

E164

Fixed Network	Country Code	Numbering Area Code	Customer Number
	44	1728	860058

Mobile Network	Country Code	Network Destination Code	Customer Number
	44	850	947867

X121

	Data Country Code	Network Digit	Network Terminal Number
	3 digits	1 digit	Up to 10 digits

Figure 3.11 Address schemes

In the E164 specification, a number is defined as a predefined country code of 1–3 digits, a numbering area code plus subscriber number of up to 14 digits, with prefix digit of 0 for national number, 00 for international number. The structure with X121 is similar, with a three-digit Data Country Code, a one-digit network identifier and a 10-digit National Terminating Number.

The overall rules for high level numbering are given by the ITU-T. Numbering plans at a national level define specific area codes as well as key nationwide service numbers (e.g. police, ambulance and fire service).

Network and Service Management

A major concern for any network operator is how to keep services up and running. The telephone network is comprised of a host of elements, all of which need to work together—exchanges, computers, communications software, transmission equipment, access lines and so on. The challenge facing the operator is to integrate the management of these separate components in a cost-effective manner.

With regard to the telephone network, management covers many different aspects, some of the more important ones being:

Service provision
The services that users require should be available on demand. It should be possible to provide and change services with minimal manual intervention. Automated processes are needed to manipulate network data so that this can be achieved.

Service Assurance
Most of the faults that occur in the network should be identified before they affect service. Network elements need to be monitored for faults or degradation of performance. Network performance should be automatically reconfigurable to restore service and fault information needs to be diagnosed, correlated, prioritised and dispatched for attention as soon as it is found.

Test and performance monitoring
Any network element, anywhere in the network, should be testable from a fault handling point. The quality of service being offered to a user also needs to be monitored in real time. A network being managed by a third party should have a service level agreement (SLA) against which acceptable levels of performance can be measured.

Inventory management
All network elements should be identifiable and their status (e.g. connected or out of service) and configuration should be known. The network

inventory should be accurate and detailed enough to ensure that it is clear exactly what constitutes the network being managed, how each element relates to another and which network elements affect which services.

Order handling
It should be possible to order/reserve resources (e.g. extra bandwidth or new services) automatically. Checks on available capacity, credit, stocks etc. should also be available as part of the network management system.

Accounting
Charging, tariffing and credit information should be maintained as part of the management system. This would be partly for information and partly for verification purposes.

There are many items that could be added to the above list, some basic and general requirements such as the security of access, others more specific, such as automated service creation.

A model, similar to the ISO 7-layer model, is used for network and service management. In this case, we have four layers—a business layer, a service layer, a network layer and underlying network elements, as shown in Figure 3.19.

Each of the network elements holds information on its configuration and well-being in a store called a Managed Information Base (MIB). These can be inspected from the Network Layer through a management interface such as SNMP or CMIP. The Service Layer converts network-centric data into customer-orientated service information and the Business Layer is concerned with the end-to-end operation of the network and its service.

Typically, the Telcos have built their own systems to monitor and control the network and its service surround. More recently, telecommunications system management has come to resemble more closely the management of distributed computing networks. This is typified in the approach to the monitoring of network elements, as shown in Figure 3.12.

The network management layer operates by polling and gathering alerts from the elements it looks after. This is done by interacting with the element's management information base, which keeps track of the status of all of the objects that constitute that element. Because the interaction is through standard protocols acting on a group of well-defined objects, it becomes possible to manage large heterogeneous networks without huge complexity.

In practice, the information that comes from the network elements needs to be processed. Typically, each will look after some part of a larger network and the real picture of what is happening can only be gleaned when other views are considered (Figure 3.13).

Events collected from a network element can be collected and correlated with those from other managed elements. This allows problems to

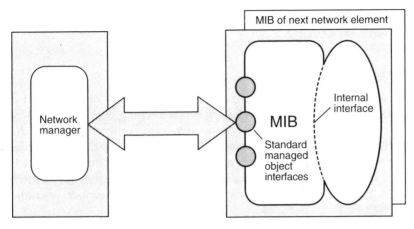

Figure 3.12 Walking the MIB

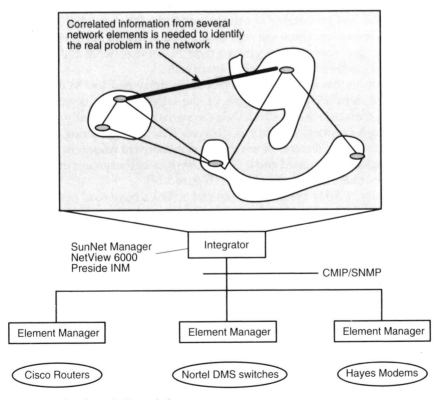

Figure 3.13 Correlation of views

be localised, for instance, a link to be declared faulty because both ends are known to be working.

Traffic engineering

This is a key issue for any telecommunications network operator as it is the basis for keeping customers happy while keeping costs down. There are inevitably some bottlenecks in any network. Some of these are deliberately inserted, as we have already seen. The key issue that needs to be addressed in traffic engineering is to ensure that the demand for service is satisfied by just the right amount of supply of resources (i.e. switch capacity, number of channels between exchanges, number of radio channels in a cellular network).

Fundamental to all traffic engineering in telecommunications is the measure of traffic intensity, known as the Erlang. This is defined as the ratio of time during which a circuit is busy to the time during which it is available. If a line is busy all the time, then the traffic intensity is 1 Erlang. Acceptable level of traffic density are defined by a telecommunications carrier on the basis of what they feel will satisfy their customer—a typical planning figure would be around 0.1 to 0.2 Erlang. This provides good service (i.e. low probability of calls being blocked) but good utilisation of network resources. There are well established tables for network capacity planning that show, for a given grade of service, what the blocking probability is for a typical distribution of calls.

Later on in this chapter we will focus on delay and loss systems. The telephone network is an example of the latter, if there is not enough capacity, then calls are blocked. Delay systems react differently, if there is not enough capacity, delivery is delayed. Traffic engineering for delay systems is quite different. When there are insufficient resources to deliver a data packet, it is queued and it is the behaviour of the queues in overload conditions that interests the planner (Figure 3.14).

Typically, a delay system behaves rather like a busy road in that hold-ups are triggered by critical overload levels and take time to clear. In

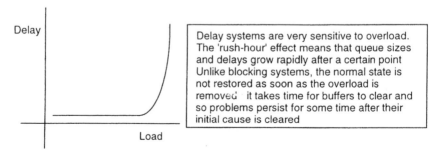

Figure 3.14 Traffic behaviour in a delay system

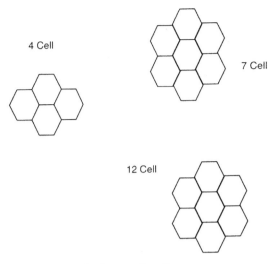

4 Cell

7 Cell

12 Cell

Figure 3.15 Cell repeat patterns

general, the smaller the queue size in the delay system, the more closely it resembles a loss system.

Mobiles

Two factors limit systems based on airborne transmission: the limited natural resource of the radio spectrum and the technology available to exploit it. Ever since the principle of frequency reuse was demonstrated at Bell Labs in the early 1980s, one of the limits was removed and the other has evaporated over the intervening years.

Cellular telephony makes efficient use of the spectrum by dividing geographic coverage into small areas (or cells) each having it own radio base station. The cells are grouped into clusters and radio channels allocated to each cluster according to a regular pattern which repeats over the coverage area (Figure 3.15).

The smaller the number of cells per cluster, the greater the number of radio channels per cell and, hence, the greater the traffic carrying capacity of that cell. There is a penalty: the less distance between clusters, the greater the interference between clusters. The ratio of repeat distance to cell radius and of signal strength to interference for each of the configurations shown in Figure 3.15 can be readily calculated.

Table 3.1

Number of cells	Repeat distance/Cell radius	Signal/interference (dB)
4	3.5	13.7
7	4.6	19.4
12	6.0	24.5

Because there is no fixed point of connection to a mobile network and in order to support roaming though cells, all mobile networks need to carry out some form of location management. Cellular networks are divided into administrative domains called location areas, which identify (approximately) which region a caller is in. The administrative domain holds the subscription records for a customer in a home database and this is used for both call connection and for accounting. Any handover between cells during a call is dealt with by call processing within the network.

There are several radio system technologies in current use. The older analogue systems include Total Access Communications System (TACS) and Advanced Mobile Phone System (AMPS) and the digital systems are Global System for Mobile (GSM) and DCS-1800. The convergence of fixed and mobile networks is being explored through an initiative known as Universal Mobile Telecommunications Service (UMTS) or Third Generation Mobile, but this has yet to impact.

3.3   DATA NETWORKS

At the beginning of the chapter, we said that we would focus on the telephone network. That was a bit of a white lie as we are now going to start looking at data networks. The reason for doing this is that the traditional telephone network is also a very capable data network—voice signals are carried as digital data and much of the world's Internet traffic is carried over the telephone network. As you might expect then, many of the issues that have been covered so far relate as much to data networks as they do to voice networks.

There are, however, some fundamental differences that separate the handling of voice and data traffic. At the heart of these differences lies the fact that voice is sensitive to delay but can tolerate error, whereas data is sensitive to error but can tolerate delay. This simple statement impacts on the strategies that have been devised in implementing data networks, so we now revisit in more detail the switching options introduced in the opening chapter.

Circuit-switching and Packet-switching

The distinguishing feature of the circuit-switching approach described above is that an end-to-end connection is set up between the communicating parties, and is maintained until the communication is complete. The Public Switched Telephone Network (PSTN) is a very familiar example of a circuit-switched network. Indeed, it is so familiar that many people are not aware that there are other ways of doing things (Figure 3.16).

Communication between computers, or between computers and termi-

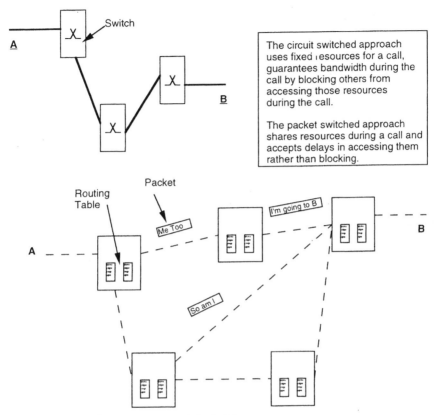

The circuit switched approach uses fixed resources for a call, guarantees bandwidth during the call by blocking others from accessing those resources during the call.

The packet switched approach shares resources during a call and accepts delays in accessing them rather than blocking.

Figure 3.16 Circuit- and Packet-Switching

nals, always involves the transfer of data in blocks rather than continuous data streams. Packet-switching exploits the fact that data blocks can be transferred between terminals without setting up an end-to-end connection through the network. Instead they are transmitted on a link-by-link basis, being stored temporarily at each switch *en route* where they queue for transmission on an appropriate outgoing link. Routeing decisions are based on addressing information contained in a header appended to the front of each data block. The term packet refers to the header plus data block.

Store-and-forward switching

The idea of store-and-forward switching in this way is older than packet-switching. It has been used in a variety of forms to provide message switched services in which users could send messages—often very long messages—with the advantages of delayed delivery and broadcast options and retransmission if the message was garbled or lost in transmission. The distinctive feature of packet-switching is that the packets, and consequently the queuing delays in the network, are short enough to allow interactive transactions.

The store-and-forward nature of packet-switching brings a number of important features that circuit-switching does not match.

Because transmission capacity is dynamically allocated to a user only when he has data to send, very efficient use can be made of transmission plant. On the circuits between packet-switches, packets from different users are interleaved to achieve high utilisation. This feature provided a strong incentive to introduce packet-switching where transmission costs are high, for instance, in North America where circuits tend to be long.

Similarly, on an access circuit to the local serving switch or exchange, the user can interleave packets for different transactions, enabling communication with numerous remote terminals simultaneously. In the early years of data communications using the PSTN-with-modems, a typical time-sharing mainframe would need a large modem room, containing perhaps hundreds of modems, to terminate the dialled-up access connections. The introduction of packet-switched services made these large modem rooms redundant overnight.

Because the network effectively buffers communicating terminals from each other they can have access circuits of different speeds. For example a large e-mail host can have high-speed network access, capable of supporting multiple simultaneous accesses, whilst a remote Personal Computer accessing the e-mail service could have much lower speed, and cheaper, network access.

Very low effective error rates can be achieved. By adding an error-detecting checksum (in effect a complex parity check) to the end of each packet errors can be detected on a link-by-link basis and corrected by retransmission. At each switching node *en route* packets would be stored until their correct receipt is acknowledged. This arrangement can also provide a high degree of protection against failures in the network since packets can be re-routed to bypass faulty switches or links without the user being aware of a problem.

Flow and congestion control

In a packet-switched network, packets compete dynamically for the network's resources (buffer storage, processing power, transmission capacity). A switch accepts a packet from a terminal largely in ignorance of what resources the network will have available to handle it. There is always the possibility therefore that a network will admit more traffic than it can actually carry with a corresponding degradation in service. Controls are therefore needed to ensure that such congestion does not arise too often and that the network recovers gracefully when it does.

The importance of effective flow and congestion control is illustrated in

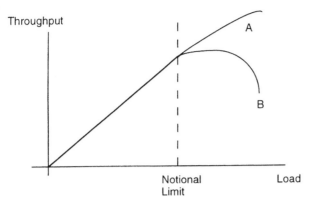

Figure 3.17 Throughput in a packet network

Figure 3.17. For a given delay performance a packet-switched network can be regarded as having a notional maximum throughput, T packets/second.

An ideal network would accept all offered traffic up to its throughput limit: beyond this some traffic would be rejected but the network's throughput would stay high (curve A). In a network with poor flow and congestion control, however, the throughput, whilst initially increasing with offered traffic, would eventually fall rapidly as a result of congestion (curve B). In practice good flow control maintains high throughput but leaves an adequate margin of safety against network failures and the imperfect operation of practical algorithms.

Performance aspects

A circuit-switched network such as the PSTN provides an end-to-end connection on demand (subject, of course, to the necessary network resources being available). Once established the users have exclusive use of the connection for the duration of the call. The connection's end-to-end delay is small (unless satellite circuits are used) and constant, and other users cannot interfere with the quality of communication. But if the required network resources are not available at call set-up time the call is refused— the caller gets engaged tone and the call is blocked. Circuit-switched networks are, in the jargon, loss systems.

In contrast, in a packet-switched network, packets queue for transmission at each switch. The cross-network delay is therefore variable, depending on the volume of other traffic encountered *en route*. Packet-switched networks are, in the jargon, delay systems. In effect a packet-switched network is a queue of queues and queuing theory is the basic tool in the network designer's toolkit. In practice the queuing problems presented by real packet networks cannot be solved analytically and performance modelling generally involves a combination of analytical and simulation methods.

Table 3.2

	Circuit-mode	Packet-mode
Delay	Short and fixed	Longer and variable
Error rate determinant	Transmission path	Packet error detection/correction
Simultaneous calls	No	Yes, packets interleaved
Speed matching	No	Intrinsic
Transmission efficiency	Low to medium	Medium to high

Table 3.2 shows the main characteristics of circuit and packet-mode services.

Connectionless and connection-orientated services

The simplest form of packet-switched service is the Connectionless or datagram mode in which each packet, or datagram, is regarded as a complete transaction in itself. There is no sense of a call being set up before communication begins and the network treats each packet independently. In the network there is no sense of packet sequence. Connectionless mode is generally used in Local Area Networks (LAN). The Switched Multi-megabit data service (SMDS), designed primarily to provide LAN-interconnection over long distances, is also a connectionless service.

Many applications, however, involve the transfer of a sequence of packets for which a connection-orientated approach is more appropriate in which a connection is established by an initial exchange of signalling packets between the communicating terminals. Data transfer then takes place and at the end of the session the connection is cleared. During the data transfer, or conversation, phase the network tries to create the illusion of a real end-to-end connection using store-and-forward switching. But to distinguish them from real connections they are referred to as virtual connections or circuits. It should be noted that circuit-mode services are intrinsically connection-orientated.

3.4   USEFUL MODELS

Models abound within the telecommunications industry. Several of these models are widely recognised and accepted. They provide a common reference point for operators, service providers and users.

The OSI reference model

This is probably the best known and most widely used model of communications. Originally developed to help eliminate incompatibility in

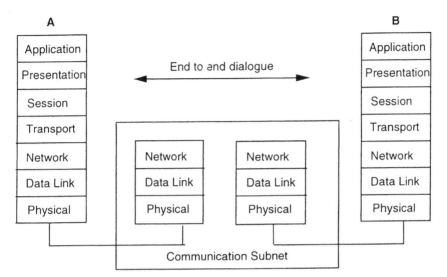

Figure 3.18 The ISO Seven Layer Model

computer systems, it has come to be used as a general model for peer-to-peer communications. In essence, it describes seven distinct layers in a communicating system. Each of these layers is specified as an international standard by ISO (Figure 3.18).

The first three layers are all about the transmission of data from A to B while, at levels four and above, the concern is with end-to-end communication. In concept, there is an interface between layers, with the higher ones passing information down to the lower ones until the physical layer is reached, whereupon actual communication takes place. At the same time that information is flowing down our segmented communicating "stack" communication takes place between peer levels in different stacks. Hence, one application will "talk" to another and both of them will be supported by six layers of underlying services.

The layers within the model are:

1. Physical—concerned with transmitting bits over a communication channel. The main issue is to ensure that, when one stack sends a '1', it is received as a '1' and not a '0'. Examples of specifications that sit within this layer are RS232-C, for electrical data signal characteristics.

2. Data Link—which builds up a frame for transmission via the physical layer. The main concern is to ensure that error free frames can be transmitted between points in the network. The ISO 3309 high level data link control (HDLC) standard and IEEE 802.3 specification for the media access control layer of an Ethernet LAN are both examples of data link protocol standards.

3. Network—which is concerned with getting an end-to-end connection, so issues such as routing and congestion control all lie here. The X121 standard mentioned earlier fits here, as does the ubiquitous Internet Protocol (IP)

4. Transport—which is the first real end to end layer and assures end to end transmission. Two (exemplary) transport layer protocols are the Internet's Transmission Control Protocol (TCP) and User Datagram Protocol (UDP). The former is connection-orientated, the latter connectionless.

5. Session—aims to ensure that different machines can establish communication. For instance, allowing two computers to complete a file transfer. The ITU-T X215 standard defines a connection-orientated session layer protocol.

6. Presentation—as the name suggests, is responsible for encoding data from a computer's internal format so that it is suitable for transmission. Hence, it has to deal with compression and decompression.

7. Application—the top of the stack, contains the communication applications that use the services of the lower layers, examples being the ITU-T X400 message handling system. This is not the application itself, but the bits within it that provide the basic communication primitives.

The seven-layer model is meant to be technology independent. It states how the communication systems should be structured, not how each part should be built. In principle, this leaves suppliers to provide components that fulfil a specific function or sit in commonly recognised part of a communicating system.

The Telecommunications Operations Map (TOM)

This contrasts with the ISO model in that it describes how a network is supported in operation rather than how it effects communication in the first place. The TOM builds on a number of previous models that describe management functions. One of these (like the ISO model) comprises a number of layers, although there are four rather than seven. This is the telecommunications management network, or TMN, model, which has four layers of management that sit over the individual managed elements that comprise the network (Figure 3.19).

As with the ISO model, each of the layers provides a capability to the layer above. The issues at the four levels are:

1. Business Management is concerned with the efficient and profitable operation of the organisation.

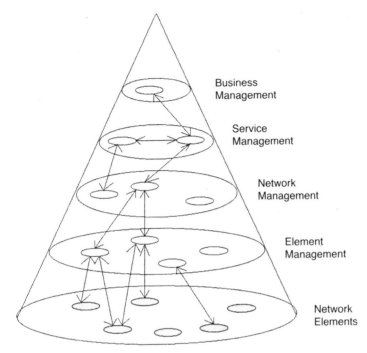

Figure 3.19 The TMN model

2. Service Management relates to the total package offered to the end user. This includes configuration, billing, performance monitoring and the like.

3. Network Management relates to the maintenance of the transmission and switching capability of the installed network. This includes fault monitoring, load balancing and the like.

4. Element Management is all about the way in which individual network components store status information, raise alarms, respond to probes and interact with the network management layer.

The TOM adds considerable detail to the basic TMN framework by describing the specific processes that have to be carried out to keep a network up and running (Figure 3.20).

In practice, these processes need to be tied together. For instance, the key management activities of Fulfilment, Assurance and Billing touch on processes at each level in the TOM. These activities overlay the operations map, as illustrated in Figure 3.21.

Part of the art of managing a network effectively is to coordinate the basic processes so that they form a consistent whole with minimal delay, overlap and margin for error. Because automation is a practical necessity

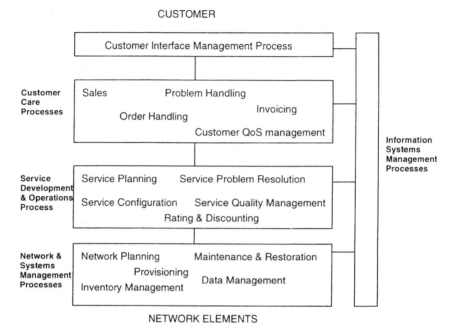

Figure 3.20   The Telecommunications Operations Map

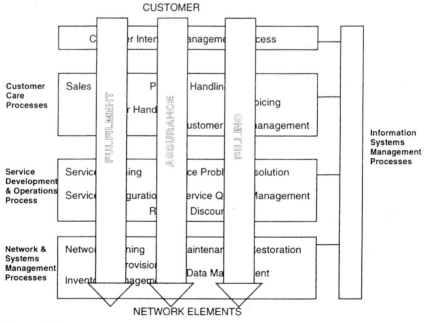

Figure 3.21   FAB activities

in achieving this, supplier equipment is judged by its fit to the TOM in much the same way that protocol standards fit into the ISO model.

3.5 NETWORK PRODUCTS

In addition to the public switched telephone network and services carried over it, there are three associated products in the current communications marketplace worth mentioning. These are:

1. Advanced telephony services. These are based on the existing telephone network and are usually provided as an overlay, either in the form of intelligent network services (such as 0800 numbers) or as customer configured partitions (such as Centrex, where part of an exchange is partitioned for use by one organisation).

2. Public Data Networks. These are special purpose services that support, for instance, the interconnection of company local area networks (LANs). Such networks are provided by a variety of service providers and they range from the simple (such as the low speed X25 packet switched service) to the advanced (such as high speed ATM technology).

3. Custom Networks. These are purpose built to meet a customer specification and usually encompass a customer's voice and data needs. They are usually built on top of the same equipment that is used for public data and telephony services and are sometimes known as Enterprise or Virtual Private Networks. A number of global alliances (such as Concert, Unisource and WorldPartners) have been established to satisfy this demanding but lucrative market.

In the remainder of this section, we illustrate each of the above with reference to a few of the many currently available products and services.

Advanced telephony services

One of the widely used advanced telephony services in the UK is BT's FeatureNet. This is an Intelligent Network product that provides premium service on the established PSTN. The service comes in two main varieties: FeatureNet 1000 and FeatureNet 500. The former is a managed virtual private network (VPN) service which closely mirrors the functionality of a private circuit network within a closed user group. The latter is a digital networked Centrex service which delivers most of the features associated with a PABX without the need for a customer to invest in capital equipment.

Services offered on FeatureNet are typical of those offered on a VPN:

- Interconnection of private exchanges, with a mix of signalling supported on the same network.

- Private numbering plan support, where customers do not need to change their existing private numbering plan on moving.

- Network Overflow, which ensures that all calls reach their destination if traffic exceeds the planned VPN capacity by using the PSTN as an overflow.

- PSTN Breakout, which enables traffic to travel as far as possible over the VPN and then out over the PSTN at local rates if there is a VPN point of presence in the charge group concerned.

The network structure that underpins FeatureNet and other similar services is based around Intelligent Network concepts, and these are explained in some detail in Chapter 7.

Public data networks

The first true data network available was based on the X25 protocol (as specified in the ITU-T/ITU X.25 standard). This standard defines an interface between a terminal (such as a PC) and a data circuit termination (which is the piece of equipment that converts data to and from packet format).

The X25 standard itself is a network access protocol. It makes no assumptions about the way in which the network itself functions, other than that the packets in any dialogue are delivered in the order in which they enter the network.

Central to the X25 is the concept of the virtual circuit (and corresponding virtual call) which enables the capacity of the carrier network to be split into logically separate channels with the packets being carried corresponding to a number of different calls being multiplexed on the same physical circuits.

Security and error-detection are built into the network and dynamic switching over various routes provides inherent protection against interception and diversion. X25 is trusted to the extent that it is used by banks to transfer billions of pounds every day. The service also is well suited for Electronic Data Interchange (EDI) order entry and sales tracking from branches and retail outlets, electronic mail, credit card authorisation.

A more recent addition to public data network offering is the Switched Multimegabit Data Service, or SMDS. Like X25, SMDS is packet based. In contrast, though, it is a connectionless service with no concept of virtual circuits: each packet is expected to find its way to its destination on its own.

The SMDS network can be regarded as a 'cloud' with a number of points of presence. Customers gain access to this 'cloud' using private circuits at typical access rates between 0.5 Mbps and 25 Mbps. Beyond this access layer, there is a core transport operating between 34 Mbps and 140 Mbps

Custom networks

In order to meet a variety of customer demands, several major suppliers have established Intelligent Network capabilities. One of the first of the global alliance, AT&T-Unisource, placed a Service Management System in Stockholm, Service Control Points in Stockholm and Amsterdam, and Service Switching Points located in Stockholm, Zurich, Madrid and Amsterdam. Customers in those four countries gain access to the Service Switching Points, and hence to a range of customised network services, via the International Switching Centre of their respective domestic PSTNs.

Access Switches, which are, in effect, Unisource Points of Presence, feeding the Service Switching Points are also located in a number of other European countries. Customers of the Unisource Virtual Private Network (VPN) in these countries gain access directly over leased lines or by PSTN switched access to the Access Switches.

The VPN Service that can be offered comprises a range of numbering, access, service and call handling features. Some or all of these can be provided, depending on the level of service required. Generally speaking, the range of such services that may be available depends on both price paid and on the perceived importance of the customer. As Intelligent Network features become better developed it will become progressively easier, quicker and cheaper to introduce features designed to meet the needs of individual customers.

It is probably apparent from the above descriptions that large-scale voice and data products are separate. There are a few instances of combined voice and data services but, in the main, each has its own technology and packaging.

3.6 SUMMARY

The telephone network is the product of nearly a hundred years of engineering innovation. As it has evolved from a novel method of communication into a global machine, specialist methods have been developed to make it work, keep it working and make it better. At the same time, this network, developed to carry voice traffic from one place to another has been used to carry data, to access the Internet, to provide remote monitoring and a variety of other things. Hence, the machine has had to be adapted to a range of different service demands and a variety of different traffic types.

The fact that all these things are possible stands as testament to the flexibility and robustness of the network and this can, in turn, be attributed to the soundness of the basic concepts used in its construction. In this chapter, we have explained these basic principles and illustrated how and where they have been applied.

Starting with the fundamental requirements of transmission, switching and signalling, we have explored the solutions used by the telecommunications engineer, where the sheer scale of the task differentiates it from many others, to provide an economical yet effective service. Practical issues such as numbering and management have been addressed and some network products introduced.

REFERENCES

Freeman, Roger (1999) *Fundamentals of Telecommunications*. John Wiley & Sons.

Freeman, Roger (1998) *Telecommunications Transmission Handbook*. John Wiley & Sons.

Minoli, Daniel (1991) *Telecommunications Technology Handbook*. Artech House.

National Computer Centre (1982) *Handbook of Data Communications*. NCC.

4
Telecommunications Technology

My life depends on the quality of the lowest bidder

John Glenn (sitting in the Mercury space capsule)

Over the last few years just about everyone will have seen a dramatic reduction in the cost of telephone calls. The whole industry has become much more competitive, more customer-focused and more dynamic. This is due, in no small part, to the availability and deployment of a wider range of technology from a greater number of suppliers.

As little as 10 years ago, it was not uncommon for a large Telco to create and build all of the technology that they needed in-house. They would defer to international standards and might use a few third parties to manufacture the equipment that they had specified, but the telecommunications network was essentially a closed shop. This has all changed with deregulation, competition and the bewildering number of mergers and acquisitions in the telecommunications sector. The industry is now much more open and there are many suppliers with some component technology to contribute. The successful telecommunications provider is the one that chooses the right technology from the marketplace and who puts it to good use.

So, in this chapter we are going to explain some of the technology that is available to the telecommunications designer. Given the sheer size of the industry, it would be impossible to do more than scratch the surface. In any case, this whole area moves so fast that many of the details that are committed to paper are out of date before it reaches its target audience.

In the next section we look in some depth at the primary mechanism for bulk transmission, SONET/SDH. This is followed by a detailed exposé of the CCITT Signalling Systems No. 7 and an explanation of the Digital Subscriber Line and ISDN technologies that promises to extend high speed data into the local network. We finish with a deeper study of some of those items already introduced, notably, satellite and radio communications and network management.

Our aim now is to use our small band of exemplars and to illuminate the key concepts of transmission, signalling and others that we have outlined but not really filled in with great detail.

4.1 BULK TRANSMISSION—SONET/SDH

Synchronous Optical NETwork, commonly known as SONET, is the TDM-based standard for high speed transmission. It is widely used by telecommunications providers to move voice traffic between switches and is increasingly being used as part of both public and private data networks. At a very simple level we could say that SONET/SDH is a structure for the signals, carried over optic fibres, that provides a high speed transmission capability.

SONET was initiated by the American National Standards Institute (ANSI), the telecommunications industry standards setting body in the United States. The standard defines optical carrier (OC) levels and electrical equivalent synchronous transport signals (STSs) for the fibre-optic-based transmission hierarchy and established the standard for connecting one fibre system to another. Following the development of the SONET standard, the CCITT (now the ITU-T) defined a synchronisation standard to address interworking between the CCITT and ANSI transmission hierarchies which is known as the Synchronous Digital Hierarchy (SDH) standard.

The basic difference between the two is that the ANSI Time Division Multiplexing (TDM) combines 24 64-kbps channels (known as DS0 signals) into one 1.544-Mbps DS1 signal, while the ITU-T TDM multiplexes 32 64-kbps channels (known as E0 signals) into one 2.048-Mbps E1 signal. With SONET built on the former, SDH on the latter, there is a disparity in rates, at least at the lower end. SONET and SDH do not actually converge until we reach a 52-Mbps base level, as shown in Table 4.1. This identifies the various SDH and SONET signals along with their speeds and capacities.

With the SONET and SDH standards defined, and the points of commonality clear, communications carriers throughout the world have a basis for the interconnection of their digital carrier and fibreoptic systems. Clearly, this is an important step forward on the previously disparate transmission standards, so let us look in a little more detail.

Table 4.1

SONET Signal	Bit Rate	SDH Signal	SONET Capacity	SDH Capacity
STS-1, OC1	51.840 Mbps	STM-0	28 DS1s or 1 DS3	21 E1s
STS-3, OC3	155.520 Mbps	STM-1	84 DS1s or 3 DS3s	63 E1s or 1 E4
STS-12, OC12	622.080 Mbps	STM-4	336 DS1s or 12 DS3s	252 E1s or 4 E4s
STS-48, OC48	2.488 Gbps	STM-16	1344 DS1s or 48 DS3s	1008 E1s or 16 E4s
STS-192, OC192	9.953 Gbps	STM-64	5376 DS1s or 192 DS3s	4032 E1s or 64 E4s

STM—Synchronous Transport Module (ITU-T)
STS—Synchronous Transport Signal (ANSI)
OC—Optical Carrier (ANSI).

SONET Payload and Transport We have seen how and why the SONET standard is based on a rate of 51.84 Mbps, termed the Synchronous Transport Signal Level 1 (STS-1). Since this is the base signal for the whole multiplexing hierarchy, the STS-1 frame structure is illustrated in Figure 4-1.

The equivalent signal to STS-1 for transmission over fibreoptical lines is termed Optical Carrier Level 1 or OC-1, which is a direct conversion from electrical to optical signalling. Higher level signals are multiples of OC-1, so OC-3 runs at three times OC-1, or 155.52 Mbps; OC-12 at 622.08 Mbps or 12 times OC-1; and so on, up to OC-192 or 10 Gbps.

An STS-1 frame consists of 90 columns and nine rows of 8-bit bytes. The first three columns are termed Transport Overhead, and are dedicated to

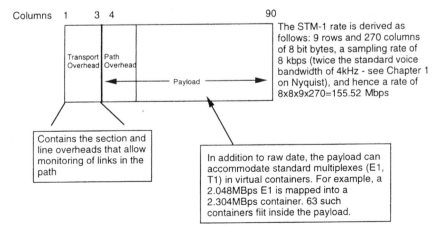

Figure 4.1   The SONET STS-1 Frame

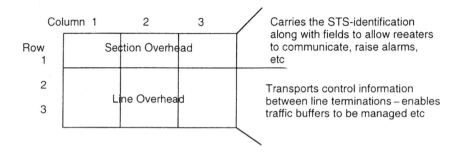

Figure 4.2 SONET Transport Overhead

network management information for the section and line parts of a SONET network as shown in Figure 4.2. The rest of the 87 columns and nine rows carry the STS-1 Synchronous Payload Envelope (SPE).

The SONET protocol overhead is significantly larger than that found in asynchronous DS1 signalling, which sends a limited amount of information in-band by robbing bits from the traffic itself. It is separated into layers that match the segments of a telephone network with the section and line parts contained in the Transport Overhead, as shown above, and the path part included with the payload.

This layered approach allows different types of equipment to be built to support different functions. The section layer defines the network segment between regenerators (the optical version of an electrical repeater). The section layer's job is to transport overhead traffic for both line and path layers as well as the actual network traffic. Framing, scrambling, error monitoring and order wiring (the capability to install or remove a service) are all done in the section layer.

The line is that portion of a network between line terminating equipment where the STS-1 signals are multiplexed to higher rates. It handles synchronization, multiplexing, automatic protective switching and additional error monitoring. In addition, it provides a data communications channel, a channel for priority installation or removal of service and room for growth. Automatic protection switching is intended to allow switching traffic from a primary to a backup circuit if the quality of the primary circuit drops below a certain threshold. Line overhead is intended to ensure that the path payload, whether data, video or voice, is reliably transported.

The path overhead is included as part of the STS-1 payload and is carried from end to end. It includes end-to-end error checking, an identifier for the type of payload being carried, and a status report on maintenance signals. The path overhead (Figure 4.3) also maps services such as DS3s into the SONET format and is accessed by terminating devices such as add/drop multiplexers.

One of the unusual features of SONET is that each overhead layer pro-

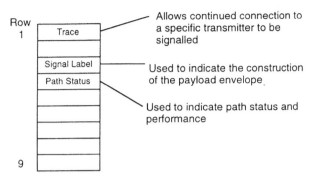

Figure 4.3 Path Overhead

tects the layer beneath it. So, if the top level signal is correctly composed and transmitted, the layers under it will also be properly delivered.

SONET uses bit-interleaved parity (BIP), a 1-byte code (consisting of 8 bits) to furnish parity for SONET frames. The Section BIP furnishes parity from regenerator to regenerator. If section parity is correctly transmitted, so should the other layers. Parity between terminating devices is achieved by line BIP, which covers line, path and traffic segments of a frame. Path BIP sets up parity between line termination equipment and covers path and traffic sections of a frame.

SONET is often deployed in the form of a set of transmission rings (Figure 4.4). This provides an alternative signal path in the event of failure; if a link is lost, data is transmitted around the 'long' part of the ring.

As well as carrying bulk traffic, nodes on the ring can deliver a low rate bit-stream to a specific location. Add/Drop multiplexers can be used insert and extract from the higher rate hierarchy. Hence SONET/SDH can be used for both core transmission and for point-to-point delivery.

Multiplexing
One of the principal benefits of SONET is that it allows for the direct multiplexing of existing network services, such as DS1, E3 and DS3, into the Synchronous Payload. So, for example with direct multiplexing, 28 DS1s or 7 DS2s are combined into the 51.84 Mbps bandwidth

Each of the lower speed signals is carried within the payload, and each can be accessed directly, without having to demultiplex the entire signal. This is a major advantage over existing asynchronous technologies, notably PDH, where signals for lower speed services are formatted differently at each step in the multiplexing chain, and the entire stream must be demultiplexed to get at the traffic and its associated overhead.

Pointers and virtual tributaries
Lower rates, such as DS1 and DS2, are mapped into the SPE by using two unique features of SONET: pointers and virtual tributaries (VTs).

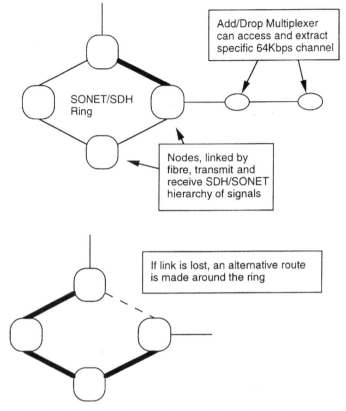

Figure 4.4 Protection in the transmission network

Traditionally, bits are inserted into incoming datastreams to accelerate them to a common data rate so they can then be multiplexed and switched. This is how four DS1s are multiplexed to a DS2, and seven DS2s to a DS3. Bit stuffing is also used to ensure that the frames of various signals are properly synchronised. The frequency difference between the clock on the receiving equipment and the arriving datastream determines how many bits are added. The additional bits are stored in buffers (sometimes known as elastic buffers). The inserted bits are introduced in random patterns so it is difficult to identify the original signal's bits from the inserted bits without demultiplexing the whole signal. Multiplexing compounds the issue, since added bits are stuffed at each level of multiplexing to compensate for various timing sources.

SONET uses pointers to do away with bit stuffing and the buffers that contain the bits, and it eliminates signal slips at multiplexers. Pointers provide easy access to actual traffic. By using pointers to mark the beginning of a frame, SONET makes it a relatively simple task to decipher a higher rate signal to locate and extract a specific DS0.

As signals go through terminals and cross-connect systems, the line and section overhead is extracted and built, based on new frame timing at each device. Different equipment at different sites will produce variation between STS signals and frame starts (even in a best synchronised network). To guard against this, an STS-1 signal could be held in a buffer until its frame coincided with the equipment that transmitted it, but this would require long buffers (up to 125 microseconds) at each step and would likely create unacceptable delays for long-distance transmission.

To overcome this, SONET lets the payload float within the STS-1 frame. A pointer is a number in the STS-1 line overhead that tells where the payload begins within the frame. If the payload is in proper position within the STS-1 signal, it can start anywhere within the frame structure and is not locked in position. If the payload shifts within the signal because of frequency variations, such as jitter or wander, the value of the pointer will be adjusted. This avoids slips and lost data that often happen with asynchronous transmissions.

Applications
SONET supports a wide range of applications, including transport for all voice services, Internet access, frame-relay access, ATM transport, cellular/PCS cell-site transport, interoffice trunking, private backbone networks, metropolitan and wide area networks, and many more. The primary benefits of SONET include:

- Applications, Service, and Protocol Transparency—since it is TDM based, there is virtually no service, protocol, or application that cannot be handled by a SONET backbone network.

- Standards—SONET is a very mature technology with standards firmly in place and being implemented by the equipment vendors, the result being that as usage increases, costs will continue to decline for the equipment itself.

- Reliability—the use of dual, self-healing fibreoptic rings, along with service restoration times in milliseconds, makes SONET one of the most reliable network backbones available.

There are four basic SONET network configurations. These are: Point-to-Point, Point-to-Multipoint, a Hub Network and a Ring Architecture. The Point-to-Point network in its simplest form can consist of two SONET (end) multiplexers linked by fibre, where the multiplexers act as concentrators of E1 and other tributaries. A Point-to-Multipoint network expands on the Point-to-Point architecture by inserting Add/Drop Multiplexers (ADMs) along the fibre link connecting the terminal multiplexers. These Add/Drop Multiplexers are devices that can pick up or drop off network traffic to other points along the fibre path (between the terminal multiplexers).

In a Hub Network, a cross connect switch serves as a central point (or hub) to interconnect multiple Point-to-Point or Point-to-Multipoint networks. Two typical implementations of a Hub Network would be to have two or more Add/Drop Multiplexers and a wideband cross connect switch to support the connection of tributary services at the tributary level or to use a broadband digital cross connect switch to allow cross connecting at either the SONET level or the tributary level.

Lastly, the most reliable and popular is the Ring Architecture, where multiple Add/Drop Multiplexers are put into a ring configuration. Typically, there would be several rings—one to cover each of the main traffic areas. Larger SONET rings follow much the same structuring principles as in the telephone network with rings connected together as a hierarchy.

The relatively simple designs of SONET networks make them easy to manage and maintain, easing the burden on network administrators. Combined with their reliability, and level of acceptance, they offer a very attractive option for all bulk transmission. Needless to say, there is an industry-led body that promotes practical deployment, in much the same way that the Frame Relay forum and ATM forum do in their areas. The SONET Interoperability Forum (SIF) was initiated by Southwestern Bell to promote agreements on standards issues in lieu of formalised standards or to be used until they are accomplished.

Now that we have a clearer picture of *the* transmission system for telecommunications, let us look at *the* signalling system that will be used to control the services that it carries.

4.2   SIGNALLING IN THE NETWORK—C7

The repertoire of signals that go between the typical end user and the Public Switched Telephone Network is fairly limited. Altogether more complex (and demanding) is the signalling between the switches that comprise the network. The universally accepted protocol for this purpose is a message-based signalling protocol known as CCITT Common Channel Signalling System Number 7, or CCSS7 (or even just C7) for short.

In its full glory C7 is a very comprehensive and necessarily complex protocol and justifies a book to itself much bigger than this one in its own right (indeed, there are several such books). Given this, we will limit ourselves here to the essentials. Figure 4.5 shows an example of the signalling involved in setting up and clearing a basic call, assuming ISDN terminals at both ends.

The calling terminal, a digital telephone say, initiates call set-up by sending an ISDN SETUP message to the originating Local Exchange. This SETUP message contains the calling and called party numbers and other information needed to establish an appropriate connection (such as

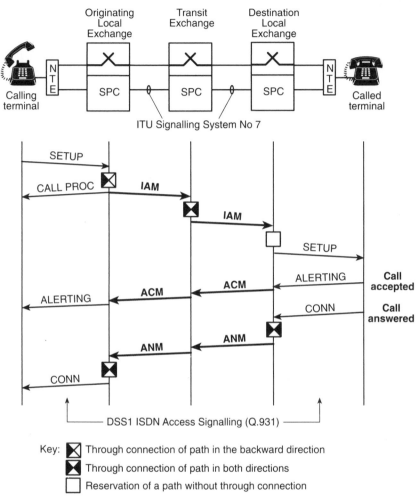

Figure 4.5 Signalling for Basic Call Set-up in the ISDN

whether a digital connection is needed from end-to-end or whether a partly analogue connection would do). The originating Local Exchange acknowledges receipt of this message by returning an ISDN CALL PROCEEDING message indicating that the network is attempting to set the call up.

The Call Control process in the originating Local Exchange then translates the ISDN SETUP message into a corresponding C7 message, which is an Initial Address Message or IAM. This Initial Address Message is routed through the signalling subnet until it reaches the Local Exchange serving the called party (the destination Local Exchange). The routing decision at each switch *en route* is based on the called party's number and any other pertinent information contained in the IAM (such as whether a satellite link is acceptable).

The destination Local Exchange translates the Initial Address Message into a corresponding ISDN SETUP message, which it delivers to the called party. The called party accepts the call by returning an ISDN ALERTING message to the destination Local Exchange. The ALERTING message is translated into a C7 Address Complete Message (ACM) which is passed back to the calling terminal as an ISDN ALERTING message as shown. The ACM both indicates to the other exchanges involved in the connection that the destination Local Exchange has received enough address information to complete the call and passes the alerting indication (i.e. that the called party is being alerted) to the originating Local Exchange.

The speech path is shown as switched through in the backward direction at the originating Local Exchange on receipt of the SETUP message and switched through in both directions at the Transit Exchange on receipt of the IAM. This allows the caller to hear any in-band signalling tones sent by the network (for a variety of reasons not all call attempts succeed).

The called telephone now rings and the originating Local Exchange sends a ringing tone to the caller. When the call is answered (i.e. the handset is lifted) the called telephone generates and sends an ISDN CONNECT message to the destination Local Exchange, which it translates into the corresponding C7 Answer message (ANM). This is passed back to the calling Local Exchange where it is translated back into an ISDN CONNECT message and passed to the calling terminal. At each switch *en route* any open switch points are operated to complete the connection in both directions, giving an end-to-end connection, and the call enters the conversation phase. Billing for the call usually starts at this point.

It is worth noting that in the case of ISDN access there is a distinction between accepting the call and answering it. The reason for this is that, unlike a PSTN access, the Basic Rate ISDN customer interface can take the form of a passive bus. This bus can support (simultaneously) a number of different terminals (up to eight), of different types (such as fax machines, telephones, Personal Computers, and so on). Hence, some calls are automatically answered, others are more like ordinary telephone calls and rely on someone picking up the phone.

The destination Local Exchange does not know until it receives the ALERTING message from the called party whether it has an appropriate terminal connected to the interface to take the call. Amongst other things, the SETUP message may carry compatibility information that the terminals may use to ensure compatibility between calling and called terminals. If there were no appropriate terminal connected to the called access the call would not be accepted.

At some later time the calling party (say) clears the call. This is signalled to the originating LE by means of an ISDN DISCONNECT message. The

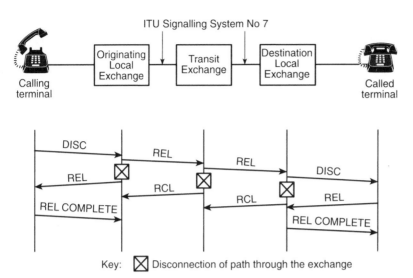

Key: ☒ Disconnection of path through the exchange

Figure 4.6 Normal Call Clear Sequence using C7

originating Local Exchange then initiates release of the ISDN access circuit by returning an ISDN RELEASE message, acknowledged on completion by the calling terminal sending a ISDN RELEASE COMPLETE message. Release of the inter-exchange circuit is signalled to the Transit Exchange by a C7 Release (REL) message, completion of which is signalled back by a C7 Release Complete (RLC) message. Successive circuit segments are released in a similar way as shown. A similar process in the other direction is used if the call is cleared by the called party. There is also the option for the called party to suspend the call for a short time by replacing the handset before resuming the call (Figure 4.6).

The basic C7 protocol stack is shown in Figure 4.7. It is a layered protocol, but was defined before the publication of the OSI Reference Model (RM) and the C7 levels of protocol, though similar, do not correspond exactly with the OSI layers. The alignment of C7 with the OSI Reference Model is a comparatively recent development.

The Message Transfer Part, or MTP, provides for the reliable, error-free transmission of signalling messages from one point in the C7 signalling subnet (referred to as a Signalling Point) to another such point. It is itself organised as three distinct functional levels, similar to but not the same as the lowest three OSI Layers.

MTP Level 1—the physical level—is usually referred to as the Signalling Data Link. It provides a physical transmission path (usually a 64 kbps time-slot in a higher-order multiplex) between adjacent Signalling Points.

MTP Level 2, usually known as Signalling Link Control, deals with the formation and sending of Message Signal Units (MSUs) over the Signalling Data Link. The functions at this level include checking for errors in

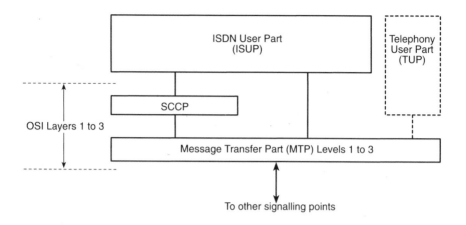

Figure 4.7   The C7 Protocol Stack for ISUP

transmission using a Cyclic Redundancy Code added to the MSU before transmission (in effect a form of parity check), and correcting any such errors by retransmitting the MSU. In this way MTP Level 2 ensures that signalling messages get neither lost nor duplicated. It also operates a flow control procedure for message units passed over the signalling link. Like Level 1, Level 2 operates only between adjacent Signalling Points. So a signalling connection between an originating and destination Local Exchange involved in setting up a call actually involves a number of independent signalling links in tandem.

MTP Level 3 is concerned with routing signalling messages to the appropriate point in the C7 signalling subnet. This is based on a unique 14-bit addresses, known as Signalling Point Codes, assigned to each such point in the signalling subnet. Routing is predetermined with alternative routes specified for use if the primary route becomes unavailable. So at each Signalling Point reached by a signalling message a decision is made as to whether the message is addressed to that Signalling Point or is to be routed onward to another. When used to route signalling messages in this way a Signalling Point is operating as a Signalling Transfer Point or STP.

The ISDN User Part, or ISUP, uses the services provided by the MTP. It is concerned with the procedures needed to provide ISDN switched services and embraces the functions, format, content and sequence of the signalling messages passed between the signalling points.

Whilst this explanation uses the ISDN case, it should be realised that the first version of C7, published in 1980, did not cover ISDN services. These were not even defined until 1984. The 1980 C7 standard defined the Telephony User Part, or TUP, which does for analogue telephone services what ISUP does for ISDN services. In practice the two ISDN and Telephony User Parts will co-exist, perhaps for many years, before TUP is entirely supplanted by ISUP. For clarity and brevity here, and because we are looking to explain the more general case, we focus on ISUP.

One of the shortcomings of the 1980 version of C7 was that signalling was defined in terms of the messages that passed between adjacent exchanges. This was fine for analogue telephony services, but the ISDN, with its powerful signalling between user and network, brought a much wider range of services into prospect. Many of these services require signalling messages to be passed between the originating and destination Local Exchanges without the intervention of intervening exchanges *en route*. Indeed, in some cases signalling is required between the Local Exchanges even in the absence of a connection being established between them.

This requirement found the MTP wanting, so, subsequently, the Signalling Connection Control Part, or SCCP, was added to C7 to provide greater flexibility in signalling message routing. Whilst the Telephony User Part (TUP) uses only the services of the MTP, the ISDN User Part (ISUP) also makes use of the SCCP. The SCCP was designed to provide the (by then) standard OSI network layer service, supporting both connectionless and connection-orientated methods of message transfer. In effect, it created a packet-switched network within the signalling subnet by means of which any Signalling Points can send signalling messages to any other Signalling Point, independent of switched connection in the switched information subnet.

4.3 EXPLOITING THE LOCAL LOOP—XDSL AND ISDN

So far we have concerned ourselves with the 'inside' of the telecommunications network. With just about every one of the established carriers having several billion dollars, euros or pounds vested in their local network (up to 30% of their assets) the final link between the local exchange and the customer's premises is a key area of concern. Such a large sunk investment cannot be overlooked, so there is considerable motivation to deploy technology that exploits what is already in the ground.

As we saw in Chapter 1, the local loop is constrained to operate at less than the Shannon capacity of a noise- and bandwidth-limited channel. So, the standard 3 kHz channel with a 30-dB signal-to-noise ratio has a theoretical data capacity of a rather paltry 30 kbps. Although perfectly satisfactory for voice traffic, the local network is somewhat wanting when it comes to data! Hence there is considerable interest in any technology that releases the local network from its analogue operating limits.

xSDL

Digital Subscriber Line (DSL) is a promising candidate in this area. It defines how a pair of modems—one located at the local telephone exchange and the other at the customer site—can be used to deliver high speed signals over their established twisted-pair copper connection. There are several varieties of DSL.

- Asymmetrical DSL (ADSL) allocates the available bandwidth in an asymmetric spectrum so that ore data is delivered downstream (toward the user) that is returned to the exchange in the upstream channel. ADSL is well suited for high-speed Internet/intranet access, video on demand and telecommuter applications. ADSL speeds range from T1 (1.544 Mbps) and E1 (2.048 Mbps) to 6 Mbps and beyond downstream. Upstream return channel speeds range from 64 Kbps to 384 Kbps to 640 Kbps. ADSL transmissions operate at distances up to 5 km (between the customer and the local exchange or serving central office switching system) via a single copper twisted-pair.

- High Speed DSL (HDSL) is a symmetric technology with speeds of 1.5 or 2.0 Mbps (upstream and downstream). Its main purpose is to replace traditional T1/E1 leased circuits. As per the standards bodies, HDSL is a two-wire implementation with an operating range somewhat more limited than that of ADSL. Over 3km telephone companies need to install signal repeaters to extend the service. Because HDSL is a two-wire implementation, it is deployed primarily for PBX network connections, digital-loop carrier systems, interexchange Points of Presence (POPs), Internet servers, and private data networks.

- Symmetric DSL (SDSL) is similar to HDSL in that it delivers 1.5 M or 2.0 Mbps (or submultiples), but it does so using a single line, downstream toward the user and upstream. The use of a single line further limits SDSL's operating range; 10000 feet is the practical limit for SDSL applications. Because of its symmetrical nature, it is well suited for videoconferencing applications or remote LAN access.

- Very High-Speed DSL (VDSL) is asymmetric. Its operating range is limited from 1000 to 4500 feet, but supports very fast transmission via single twisted-pair copper. Data can travel at rates up to 51.84 Mbps from 330 to 1000 feet with rates of up to 1.6 Mbps on the upstream return path. VDSL is positioned as the eventual modem of choice for fibre-based full-service networks. The extra bandwidth allows telephone companies to deliver High-Definition Television (HDTV) programming using VDSL technology.

- Rate Adaptive DSL (RADSL), which automatically adjusts to copper quality degradation or can be manually adjusted to run at different speeds up to ADSL rates.

- IDSL or ISDN-based DSL, which inverse multiplexes two ISDN 64 bps B channels using 2B1Q coding into one 128 kbps channel.

All of the DSL technologies have been subject to some work within the ANSI, the body that sets US standards for telephone line transmission, and ETSI (European Telecommunications Standards Institute), its European counterpart. The ITU-T (formerly CCITT), the world-wide telecommunications standards body, has not yet addressed DSL because the systems are intended for local network services and so do not generally cross national boundaries.

In terms of maturity, ADSL is the most advanced. Implementations of ADSL are based on three types of line modulation scheme: 2B1Q, Carrierless Amplitude Modulation (CAP) and Discrete Multi-Tone (DMT)

2B1Q

This is the ISDN option and it converts blocks of two consecutive signal bits into a single four-level pulse for transmission. Therefore, the information rate is double the baud rate as there are two bits per baud arranged as one quaternary (four-level) pulse amplitude modulated signal. All four possible combinations of two information bits map into a quaternary symbol. This is also known as 4-Pulse Amplitude Modulation (4-PAM). The mapping from pairs of bits to voltage levels is given in Table 4.2.

Table 4.2

Bits	Transmit Level
10	+3V,
11	+1V
	−1V
00	−3V

CAP

Carrierless Amplitude/Phase modulation is a legacy non-standard technique that uses two-dimensional eight-state Trellis coding, Viterbi decoding, and Reed-Solomon forward error correction. Adaptive channel equalization is used to support required performance in the presence of channel impairments. This technique is relatively inexpensive and technologically proven. It is used to support 1.5 Mbps downstream and 64 kbps upstream. CAP uses the frequency above 4 kHz with the lower frequency band being used for ordinary telephony.

DMT

Discrete Multitone Modulation is a multicarrier method that subdivides the information channels (telephony, upstream data and downstream data) into 256 subchannels onto which the traffic is mapped. DMT is the ANSI ADSL standard. It enables the hardware to distinguish and to isolate subchannels that are not performing as expected (e.g. high noise and high attenuation) and to assign the traffic to a neighbouring subchannel. The low-rate carrier frequencies are Quadrature Amplitude Modulated. In ADSL, it supports 6 Mbps upstream and 640 kbps downstream.

With local loop generally one of the bottlenecks in supporting advanced applications, there are a number of technologies that seek to extend its capacity. As well as xDSL, there are also technologies such as ISDN which exploit the existing copper links along with Passive Optical Networks and short-range radio that provide a solution to the bottleneck with an alternative medium (more on these later).

Whatever technology is used, the upgrade to higher speed transmission systems is not achieved as easily in the local loop as in the long haul network (e.g. deployment of SONET/SDH) because of the number of facilities involved. For this reason, the local loop is often seen as a problem by both the carriers and by other interested carriers (in particular, Internet Service Providers or ISPs).

ISDN

The Integrated Services Digital Network, introduced in Chapter 1, is an extension of the public switched telephone network that replaces the analogue local loop with a digital one. It was designed around the notion of separate channels operating at 64 kbps, this number springing from the fact that basic, analogue voice transmission requires 8 k samples per second, each of which is encoded as 8 bits.

In the UK and Europe, ISDN is offered in two forms, ISDN2 and ISDN30, where the number suffix denotes the number of 64 k channels that are provided. ISDN2, also known as Basic Rate Access, gives you two 64 k (B or bearer) channels and a single 16k signalling (D or delta) channel. ISDN30 is also called Primary Rate Access and provides 30 B channels along with a D channel. In the USA, Primary Rate Access is based around 24 B channels, with one D channel. In both cases, Basic Rate is intended for home use, and Primary Rate is meant for businesses.

In practice, there are many applications that require some number of channels between 2 and 30. High quality videoconferencing, for example, requires around 6 B channels. There are several approaches to getting the right speed to suit a wide variety of services, a technique known as inverse multiplexing. The most common method (called BonDing—Bandwidth on Demand Interoperability Group) can be used along with standard ISDN channels to support up to 63 combined B channels. Other options include Multilink PPP (designed for Internet traffic over ISDN) and Multirate Service (an Nx64 service, provided as part of the ISDN service).

By way of contrast, there are times when you want to use a 64 k channel to carry data of a lower speed. This is known as rate adaption and, again, there are standards for this. The two most common are known as V.110

and V.120. Both ensure that slower data is carried safely over the higher-speed bearer.

The ISDN equivalent of the telephone socket, at least in the UK, is called the Network Termination Unit or NT1. This is a box that has copper wires going back to the main telephone network on one side and a socket, like the standard phone socket only a bit wider, on the other. ISDN compatible equipment plugs directly into the NT1. If not you need a terminal adapter or TA which is used to connect ISDN channels to the interfaces you have on most current computing and communications equipment (i.e. RS-232 and V.35).

The precise details of where and how you connect varies from place to place. ISDN standards use a set of 'reference points' (such as the S/T interface between the NT1 and the TA) as a basis for interworking between devices and to define the boundary between the phone network and your private installation.

Up to eight devices can be attached to one ISDN line and these can be placed anywhere on a 'bus' connected to the S/T point but there are limits on how far this bus stretches (typically about 200 m).

Transmission over the local network with ISDN is based on a technique known as echo cancellation. This provides full-duplex operation over the two-wire subscriber loop and works by keeping a record of transmitted signals so any interference (or crosstalk) between sent and received signals is removed.

The system is intended for service on twisted-pair cables for operation to around 5.5 km. Echo cancellation can be contrasted with time compression multiplexing (TCM), another method that has also been used for local network transmission. In TCM (also called the ping-pong method), data is transmitted in one direction at a time, with transmission alternating between the two directions. To achieve the stated data rate, the subscriber's bitstream is divided into equal segments, compressed in time to a higher transmission rate. These segments are transmitted in bursts which are expanded at the other end of the link to the original data rate (the actual data rate on the line must be more than twice the data rate required by the user). A short guardband period is used between bursts going in opposite directions to allow the line to settle down. To use this technique in ISDN, which has a total user data rate of 144 kbps, requires over 288 kbps to be transmitted over the copper loop.

With echo cancellation, digital transmission occurs in both directions within the same bandwidth simultaneously. Both transmitter and receiver are connected through a device (known as a hybrid) that allows signals to pass in both directions at the same time. The problem here is that echo is generated by the reflection of the transmitted signal back to the user. Both near-end and far-end echo occurs. The near-end echo arises between the sender's hybrid and the cable; the far-end echo is from the receiver's hybrid device. The magnitude of the echo is such that it cannot be ignored and the technique used to overcome this problem is echo cancellation. This works by calculating the composite echo signal and subtracting this from the incoming signal, thus restoring the true signal.

In order to provide an economic Basic Rate service, the digital loop must be implemented without conditioning the plant, special engineering or special operations. The selection of a appropriate line code is a key part of meeting this requirement as it determines both the transmission characteristics of the received signal and the added near-end crosstalk noise levels on other pairs in a multipair cable.

The ISDN Basic Rate line code specified for North America is 2B1Q (2 binary, 1 quaternary). Introduced earlier, this is a four-level pulse amplitude modulation (PAM) code without redundancy that groups data into pairs of bits for conversion to quaternary symbols. In each of these pairs of bits, the first is called the sign bit and the second is called the magnitude bit.

2B1Q line code was selected to support ISDN transmission for local loops of up to 5 km. The ISDN signal is transmitted in full-duplex mode, bidirectionally on the same pair of wires. As noted, to accomplish this, ISDN transceivers must contain a hybrid function to separate the two directions of transmission. As we have seen, for the receiver to differentiate between far-end transmission and reflections from the near-end transmission, echo cancellation techniques are used. The operating range of ISDN is dictated by the amount of attenuation and near-end crosstalk (NEXT) from adjacent 2B1Q ISDN signals.

In 2B1Q, each successive pair of scrambled bits in the binary datastream is converted to a quaternary symbol to be output from the transmitter at the interface. At the receiver, each quaternary symbol is converted back into a pair of and then reformed into the bitstream that represents the original B and D channels.

ISDN, like xDSL, is a technology that seeks to exploit the local loop by getting more bits down the line. For all their difference, they have a lot in common. Indeed, one of the many varieties of xDSL (IDSL) brings the two directly together.

Another approach to the 'last mile' problem (i.e. how to avoid the bottleneck of the local loop) is to replace the copper link with a more flexible and higher capacity alternative. In the vanguard of this is the radio technology that we now focus on.

4.4 SATELLITE COMMUNICATIONS

A communication satellite is an overhead wireless repeater station that provides a microwave link between two geographically remote sites. Because of its high altitude, a satellite can transmit via a wide area over the Earth's surface. In addition to various 'transponders', each satellite consists of a transceiver and an antenna tuned to a certain part of the allocated spectrum. The incoming signal is amplified and then rebroadcast on a different frequency. Most satellites simply broadcast what they receive and have mostly been used to support applications such as TV broadcasts and voice telephony. Satellites are increasingly playing their part in Wide Area Networks (WANs), where they provide backbone links to geographically dispersed Local Area Networks (LANs) and Metropolitan Area Networks (MANs). Satellite channels are characterised by a wide-area coverage of the Earth's surface with consequent transmission delays and by a large channel bandwidth and transmission costs largely independent of distance

Satellite links operate in different frequency bands and use separate carrier frequencies for the uplink and downlink. The use of C bands, the most commonly used in first-generation satellite networks, is already crowded since terrestrial microwave links also use these frequencies. Although the use of higher frequencies of Ku and Ka bands is becoming the trend, attenuation due to rain is a major problem in both of these bands. Also, due to the higher frequencies, microwave equipment is still very costly, especially in the Ka band (Table 4.3).

Modern satellites are often equipped with multiple transponders. The area of the Earth's surface covered by a satellite's transmission beam is referred to as the footprint of the satellite transponders. The uplink is a highly directional, point-to-point link that uses a high-gain dish antenna at the ground station. The downlink, having a large footprint, provides coverage for a substantial area or a spot beam to focus high power on a small region. This results in less costly and smaller ground stations. Some satellites can dynamically redirect their beams and thus change their coverage area.

Satellites are positioned in orbits with different heights and shapes (circular or elliptical). Based on the orbital radius, all satellites fall into one of three categories: Low Earth Orbit (LEO), Medium Earth Orbit (MEO) or Geostationary Earth Orbit (GEO). The issues and characteristics in each case are shown in Table 4.4. In addition to their orbit characteristics, satellites are also classified in terms of payload. Satellites that weigh in the

Table 4.3

Band	Uplink (GHz)	Downlink (GHz)	Issues
C	4 (3.7–4.2)	6 (5.925–6.425)	Interference with ground links
Ku	11 (11.7–12.2)	14 (14.0–14.5)	Attenuation due to rain
Ka	20 (17.7–21.7)	30 (27.5–30.5)	Attenuation and high equipment cost
L/S	1.6 (1.610–1.625)	2.4 (2.483–2.500)	Interference with ISM band

Table 4.4

	LEO	MEO	GEO
Height	100–300 miles	6 k–12 kmiles	20 kmiles
Time in line of sight	>15 min	2-4 h	24 h
Strengths	Lower launch costs, very short round-trip delay, small path loss	Moderate launch cost small round-trip delays	Covers much of the Earth's surface
Limitations	Very short life, Encounters radiation belts	Larger delays than LEO, greater path loss	Long round-trip delay costly Earth station

range of 800–1000 kg are termed small, while the heavier satellites are considered big. While GEO satellites are typically big satellites, MEO and LEO satellites can fall in either class.

Satellite channels have some unique characteristics that require special considerations for data link control (layer 2 of the OSI model). This is because satellite links have both high bandwidth and high delay. Also, since they typically provide a broadcast channel, media-sharing methods are needed at the Media Access Control (MAC) sublayer of the data link layer. The Carrier Sense Multiple Access/Collision Detection (CSMA/CD) schemes, used in Local Area Networks (LANs), such as Ethernet cannot be used with satellite channels. This is because it is not possible for Earth stations to do carrier sense on the uplink (due to the point-to-point nature of the link). And a carrier sense at the downlink informs the Earth stations about potential collisions that might have happened 270 ms ago. Such delays are not practical for implementing CSMA/CD protocols.

Most satellite MAC schemes assign dedicated channels in time and/or frequency for each user. This is because the delay associated in detecting and resolving multiple collisions on a satellite link is usually unacceptable for most applications.

Typical MAC schemes are:

- Frequency Division Multiple Access (FDMA), the oldest and still one of the most common schemes for channel allocation, uses the available bandwidth, broken into frequency bands, for different Earth stations. This requires guard bands to provide separation between the bands. Also, the Earth stations must be carefully power-controlled to prevent the microwave power from spilling into the bands for the other channels.

- Time Division Multiple Access (TDMA), a scheme that sequentially time-multiplexes channels. Each Earth station transmits in a fixed time slot only. More than one time slot can be assigned to stations with more bandwidth requirements. This requires time synchronization between the Earth stations, which is generated by one of the Earth stations and broadcast via satellite.

- Code Division Multiple Access (CDMA), which uses a hybrid of time/frequency multiplexing. CDMA is a form of spread-spectrum modulation that provides a decentralised way of providing separate channels without timing synchronisation. It is a relatively new scheme but is expected to be more common in future satellites.

In general, channel allocation can be either static or dynamic. The latter is becoming increasingly popular with Demand Assignment Multiple Access (DAMA) systems allowing the number of channels, at any time, to be less than the number of potential users. This is much the same as explained earlier for the paths though an exchange being less than the number of connected customers.

Satellite connections are established and dropped only when traffic demands. Long round-trip is a major issue that affects the design of data link control layers in satellite data networks. In order to isolate the effect from higher layers and provide transparent interface with other terrestrial networks, most schemes use a combination of different protocols to achieve higher channel efficiency and make channel allocation highly dynamic.

A TDMA satellite channel consists of multiple time slots in a framed structure. Each time slot carries data packets belonging to one of many protocol users. The assignment of time slots to a channel is not fixed but supported dynamically in real-time. Each data packet carries a Virtual Circuit Identifier (VCI) field that indicates its receiving Earth station. Different Earth stations can recognise their packets on the downlink broadcast by checking the VCI field in the packet.

At any instant, a TDMA frame consists of fixed numbers of reserved and empty slots. This information is broadcast to all Earth stations via satellite broadcasts. When an Earth station has reserved a slot, the packets are termed as safe packets.

Whenever a new Earth station tries to establish a channel, it sends an unsafe data packet in one of the available slots. If the packet reaches the destination without collision, the slot is reserved and the Earth station can proceed to transmit safe packets. If a collision occurs, an avoidance scheme is used to resolve the conflicts between the contenders. A user that has no data to send in a reserved time slot loses reservation for that slot. A user can reserve more than one time slot if required.

The application of TDMA to a satellite environment depends on the round-trip delay before an Earth station can transmit safe packets (can be several hundred ms), the application and its required Quality of Service (QoS) and the Bit Error Rate (BER) of the link.

A LEO satellite environment, for example, has small delays and low

BER and is thus well suited to packetised voice applications. For the same application, however, a GEO satellite link is not well-suited mostly due to longer delays.

An alternative to TDMA is Code Division Multiple Access (CDMA), a type of spread-spectrum communication, initially used only in military satellites to overcome jamming and provide security to the user. Each binary transmission symbol is represented by a spreading code consisting of a zero/one sequence. The bit rate of the code is typically much higher than the symbol bit rate. Each user has a unique code that is orthogonal to all the other codes. The resulting signal is obtained from the product of the input datastream and the spreading code. At the receiver, the incoming bitstream is correlated with the receiver's spreading code, and user data is retrieved. If the data is not meant for that user, because it has a different spreading code, it appears as noise. The CDMA process method of modulation has several advantages. First, it solves the problem of multiple access without any coordination among the users. Each user can transmit its data at any time without interference from the other user. Second, the spreading code provides a method to identify and authenticate the source transmitter Earth station without explicit information in the packet. It also improves security, as it is difficult to detect a user's pseudo-random spreading code. Lastly, the method allows the reuse of same frequencies in adjacent beams in a multiple spot beam satellite by assigning different spreading codes to each user.

The main limitation is that, as the number of users increase, the error rate degrades, making it extremely difficult accurately to determine the degradation in system performance with increasing simultaneous users. Still, due to its apparent advantages, most modern satellite systems are employing CDMA as the channel access method.

Very Small Aperture Terminals (VSATs) are the most common 'other end' of the satellite link. Also known as micro Earth stations or personal Earth stations, VSATs allows for reliable transmission of data via satellite using small diameter antennas of typically 0.9–1.8 m. With their great reliability, versatility, and flexibility, the technology offers a cost-effective alternative to other communication options.

Both C-Band (4 GHz/6 GHz) and Ku-Band (11 GHz/14 GHz) are used in one-way VSAT networks. Declining Ku-Band equipment costs and crowding of the C-Band by terrestrial microwave radio systems have increased the popularity of Ku-Band systems in recent years. More popular today is the two-way or interactive configuration capable of handling both voice and data. Interactive networks offer a fast solution of providing reliable data communications in environments of embryonic telecommunications infrastructures and are being used extensively in developing countries throughout the world and in Eastern Europe. Depending on the geographic location, VSAT antennas usually range from 1.2m to 2.4 m.

This central host serves a number of geographically dispersed terminals organised as a star around the host (e.g. a central office with branches all over the country).

The most common MAC schemes used on VSATs is TDMA. A window-based protocol with selective reject Automatic Repeat reQuest (ARQ) retransmission strategy is used. The most common implementation uses a transmission window with N=128 packets, and the receiver responds with retransmission requests for only erroneous or missing packets. This protocol combined with Forward Error Correction (FEC) produces reliable data transfers combined with low average delays on satellite links. The most commonly used network protocol on VSAT links is X.25 as it was designed for an unreliable medium (and, for the same reason IP is not so well suited here).

Before moving on, we should make some mention of microwave communications—point-to-point radio operating across the frequencies of 1 GHz to 30 GHz of the electromagnetic spectrum. Once the exclusive territory of the common carriers, microwave communication has become a major competitor of standard, wireline telephone communications. Generally speaking, microwave communication involves sending waves of information between a radio transmitter and a radio receiver, each mounted on a tower. As microwave communication requires a clear 'line of sight', there must be no obstructions in the path followed by the waves, so the towers must be high enough to avoid interference from buildings, trees, etc.

Today, there are a number of different mobile radio systems ranging from pagers to the pan-European digital cellular radio system (GSM). On the horizon is Universal Mobile Telephony System (UMTS)—a wideband CDMA wireless technology.

The next big step in mobile satellite communications will be the introduction of handheld satellite telephone systems. While INMARSAT has plans to introduce its small, lightweight P-phone global system, there are many other contenders in the market, some using geostationary satellites and some using low earth orbit (LEO) satellites. Among the advocates of low earth orbit (LEO) systems, Motorola's global personal communications project Iridium is the one of the more prominent. The IRIDIUM system is a satellite-based, wireless personal communications network designed to permit any type of telephone device to communicate with any other type of telephone device. The system supports voice, paging, fax, or data.

IRIDIUM uses L-band frequencies (1616 MHz–1626 MHz) for telephone communications. For intersatellite links, gateways, and feeder link connections, Ka-band frequencies (23.18 GHz–23.38 GHz) are used. In addition, the 19.4 GHz to 19.6 GHz frequency band is used for uplink, and the 29.1 GHz to 29.3 GHz frequency band for downlinks.

4.5   NETWORK AND SERVICE MANAGEMENT

It is one thing to build a network, another to get the most from the time and money that you have invested in it. To paraphrase Churchill, the delivery of a network is not the end or even the beginning of the end. It is the merely the end of the beginning.

The effectiveness of any network depends on how well it performs—day after day, month after month, year after year. And that means both maintaining the equipment and ensuring its continued relevance. The former is usually called network management, the latter is the essence of service management. A working definition for each would be that

- Service Management relates to the total package offered to the end user. This includes configuration, billing, performance monitoring and the like.

- Network Management relates to the maintenance of the transmission and switching capability of the installed network. This includes fault monitoring, load balancing and the like.

Both of these fit within the TMN model of communications management, illustrated in Figure 4.8. At the top of the model comes the business

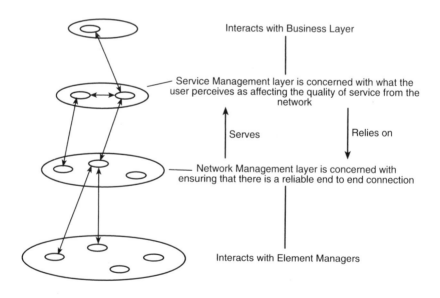

Figure 4.8   The TMN communications management model

that the network is supporting, at the bottom, the various elements of the network. In between are the various management responsibilities.

The scope of activities in the central two layers of the above figure have been described and categorised by the ITU-T (in the X700 recommendation) and ISO, as follows

- Fault   Identifying and recovering from faults

- Configuration   Keeping an up to date inventory of network resources and controlling changes

- Accounting   Managing the financial aspects of the network , such as charging and asset auditing.

- Performance   Measuring and changing performance parameters

- Security   Ensuring that the network can be accessed and used only by authorised users

To see how we go about fulfilling these management requirements, we have to build on a basic concept—that of object orientation. This concept (which is explained in some detail in the next chapter) has been in practical use on the public telecommunications network for many years and with great success. A network can be thought of as being comprised of many individual elements, or objects, that act independently. Messages relay instructions back and forth among these elements telling them how to behave for each call.

With reference back to an earlier topic, the messages between network elements are carried over a dedicated facility—our old friend, Signalling System No. 7. The behaviours that result from this messaging include setup, call proceeding, connect, release and disconnect. Each network element responds appropriately to these messages and performs actions based on a set of embedded switching methods.

The network's object orientated basis permits relatively easy expansion of the geographical reach of a network, upgrades of software for each network element, and enhancement of customer services without affecting the performance of the entire switching fabric and without having to take the network offline to isolate and correct problems.

In the networking environment, the collection of object definitions with which a given management system can work is called the Management Information Base (MIB).

The MIB is a repository of information necessary to manage each of these various network devices. The MIB contains a description of objects on the network and the kind of management information they provide. These objects can be hardware, software, or logical associations, such as a connection or virtual circuit. An object's attributes might include such things as the number of packets sent, routing table entries, and protocol-specific variables for IP routing.

- Attributes   An object is defined by properties called attributes. In the case of the network elements mentioned previously, the attributes include such things as the object's condition (on or off) and the kind of notifications it emits, such as alarms and performance statistics. Printer attributes are usually simpler, consisting of offline or out-of-paper notifications.

- Class   In object orientation, the definition of object types is separate from the actual objects. The definition—or class—is a template for the object, determining what it looks like, how it behaves, and how it is accessed. Each object is created from a class and is called an instance of that class. The process of creating an object is called instantiation. A new class can be created from an existing class. Objects that are formed from the new class are essentially the same as those created from the original, having the same data and behaving in the same way, except for any difference specified in the new class definition.

- Class Library   A collection of reusable software components is called a class library. Class libraries make it possible for developers to construct object-orientated applications by picking and choosing what they need.

- Methods   While objects are the things that are manipulated, methods describe what is to be done with the objects. Objects can be handled in a number of different ways. On a computer screen, for example, an icon that is associated with a set of files can be grabbed by pointing to it with the mouse, dragged to another icon of a trash can, and dropped into the trash can. This is the way Macintosh users erase their files. In Norton Desktop for Windows, the same thing is done using the shredder icon. In much the same way, dragging the file folder icon to the printer icon and dropping it there causes the contents of the file folder to be printed.

- Messaging   Objects are requested to perform their functions through messaging. On a LAN, for example, the client object sends a message to the server object. The message consists of the identity of the server object, the name of the function to be performed and, in some cases, optional parameters such as passwords and other privileges. The functions are limited to those defined for that object.

Management Protocols

Now that we have a means of systematically defining and reasoning about network elements, we need to see how they work with each other. And this means defining some standards for how they talk to each other— standard management protocols. There are two major options.

Simple Network Management Protocol
SNMP evolved within the Internet community as a de facto standard for managing bridges, routers, and similar communication equipment in

TCP/IP networks. Developed by several network managers responsible for running portions of the Internet, SNMP is a simple, lightweight management protocol that is easily implemented although somewhat limited in functionality. SNMP accomplishes the following three design goals:

1. Minimise the number and complexity of management functions executed in agents.

2. Provide easily expandable functions for monitoring and controlling network operation and other management aspects.

3. Maintain an architecture independent from particular hosts or gateways.

More recently, the SNMP management model has been extended to encompass systems management as well. Technology available today for SNMP-based systems management includes implementations modifying SNMP's master/agent architecture, as well as the definition of new Management Information Bases for systems management. The impending availability of new transport mappings in SNMPv3 (inherited from SNMPv2) will further enhance the protocol's viability for systems management.

SNMP does not specify which data, objects, or variables are used for network management or how network management information is represented. Instead, it uses the MIB definitions for this information.

SNMP agents support the following five Protocol Data Units (PDUs):

1. Get Request—request to get MIB variables.

2. GetNext Request—request to interrogate a table without specific entry names.

3. Get Response—answer to a Get, GetNext, or Set request.

4. Set Request—request to change a MIB value.

5. Trap—notification of a certain event containing limited data describing a problem.

The fundamental command operations of SNMP are Set and Get. The protocol issues a request to set or get a list of MIB attributes using the SNMP syntax of Objectname. The GetResponse command is the response to both the Set and Get. GetNext extends the Get command to multiple instances.

SNMP defines a limited set of traps to indicate major problems in the network. It is intended that most information about the network's health will be acquired by polling devices. The Trap PDU actually redirects polling to critical network areas.

Common Management Information Protocol

Both systems and network management were considered part of ISO standards from the very beginning. CMIP supports an object-orientated request/multiple-reply exchange of management data between managing and managed systems. The OSI model provides a large set of management capabilities and provides mechanisms to govern the amount of traffic generated by management functions. In addition, CMIP distinguishes between operations on objects and operations on attributes of objects.

A client/server model is assumed in which the client is the managing system and the server is the managed system. The managed system (server) assumes an agent role, receiving management notifications, executing commands, and forwarding event notifications. The managing system (client) invokes operations and receives notifications. Common Management Information Services (CMIS) defines services, and Common Management Information Protocol (CMIP) defines the information transfer mechanism. The structure allows different protocols to support the same services.

One of the major differences between SNMP and CMIP is the former's reliance on polling. In an SNMP environment, managed devices are regularly polled for information of interest. Agent devices then return the requested information. The managing system is responsible for requesting and filtering information. This approach can create large amounts of network traffic and imposes a significant processing penalty on both the manager and the agent. Polling is a substantial overhead in larger networks as the number of managed devices increases. In these networks, polling cycles become long and unresponsive.

In contrast, CMIP adopts an event-driven view of the world. Event reports are generated to report important spontaneous events. Event reports also supply performance information. In this way network overhead is considerably reduced, thus expanding the scalability of network management systems. The event-driven approach of CMIP more closely models the uncertain nature of network faults. In wide area networks where transmission costs are at a premium, CMIP's event-driven solution that off-loads the network by distributing the burden between manager and agent is more cost-effective than SNMP's reliance on polling. In smaller networks, however, the simplicity and cost-effectiveness of SNMP usually prevails.

4.6 SPEECH CODING

For all of the hype that surrounds the Internet and predictions that a datawave is about to hit, voice communication remains the basic fodder of telecommunications and the service central to most organisations and for residential users. Up to the early 1960s, virtually all voice was based on the transmission of a 3–4 kHz analogue signal. For close to 40 years now, voice has been carried in ever growing proportions in digital format. For most of that time, this has entailed the sampling of the voice signal 8000 times a second (at least once every analogue half-cycle) and then coding each sample using 8 bits to give a 64 kbps stream to represent the voice (or 56 kbps in the USA).

The deployment of compressed digital voice as a commercial concern is fairly recent. In combination with the carriage of voice over data networks, it offers possibilities of considerable statistical gains. And with advances in the enterprise networking industry, expanded use of the Internet, the introduction of intranets and the corporate deployment of Asynchronous Transfer Mode compressed voice is attracting considerable interest

Sophisticated voice coding methods have enabled by the evolution of VLSI technology. Coding rates of 32 kbps, 16 kbps down to 4 kbps and even less, have evolved (albeit at the cost of speech quality). The obvious implication is that one can double or quadruple the voice-carrying capacity of an established network without the introduction of new transmission equipment.

There are two classes of methods to digitize voice: waveform coding and vocoding. With the former one attempts to code and then reproduce the analogue voice curve by modelling its physical shape. The number of bits per second needed to represent the voice with this method is around 16 kbps, down to 9.6 kbps, depending on the technology. Vocoding attempts to reproduce the analogue voice curve by performing a mathematical analysis. It transmits is a small set of parameters that describe the nature of the voice curve. The number of bits per second to represent the voice with this method can be very low—9.6 kbps down to as little as 1200 bps, depending on the technology. Not surprisingly, voice quality is increasingly degraded as the digitisation rate is reduced.

There is a variety of speech coding techniques now available. The most commonly used voice digitisation/compression algorithms are:

- Pulse Code Modulation/Adaptive Differential Pulse Code Modulation (the method used traditional telephone network that works at rates of 64 kbps and 56 kbps down to 32 kbps).

- Code Excited Linear Predication/Algebraic Code Excited Linear Predication (CELP/ACELP) (that have been used over Local Area Networks and the Internet to give speech at 16 kbps and below).

- Proprietary pre-standards methods, such as Adaptive Transform Coding/Improved Multiband Excitation.

The ITU-T has produced standards for the first two forms of speech encoding—G.711 for PCM and G729 for CELP. Issues such as how to deal

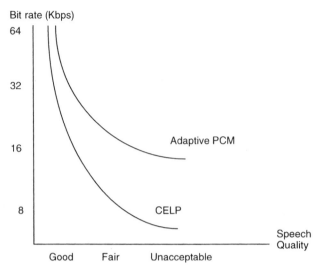

Figure 4.9   The quality of speech at low bit rates

with silence detection are still being addressed but compressed speech is now being deployed. The trade off between speech quality and bandwidth used is well quantified and shown in Figure 4.9.

Of course, when 2, 4 or even 8-fold increases are seen in the effective capacity of a telecommunications network, there is significant impact on the dimensioning sums that determine the trade off between how much is spent on switches etc and the grade of service that is offered. Having mentioned grade of service, we take the remainder of this chapter to explain this and outline some related concepts.

4.7   TRAFFIC ENGINEERING

This is a part of the operational design of a telecommunications network for determining its capability to handle a theoretical load. As such it deals with capacity, throughput and the grade of service that can be offered. The discipline is stable in the sense that the fundamental techniques have been around for nearly 80 years. Changes are afoot, though, mostly due to the spread of data networking applications.

For voice networks, traffic engineering is used for configuring communications switches and estimating the number of circuits required by a network. Traffic engineering formulas are based on probability theory, applied to predicting whether calls (traffic) will be blocked or completed through a network. Traffic components are traffic intensity, grade of service, and the busy hour. And these concepts can be manipulated using established traffic formulas, classified as loss formulas and delay formulas, to exercise a network model.

Traffic Intensity

This refers to the number of active calls in the system in addition to people demanding service from the system. Within telephony, traffic intensity is the number of calls handled by a switch or a pure voice network. From a statistics viewpoint, traffic intensity is defined as the mean service time for customers divided by the mean time between customer arrivals. Telephone engineers use Erlangs, named after an early twentieth-century Danish mathematician and teletraffic pioneer, to measure traffic intensity within a system or any traffic-dependent unit in telecommunications. An Erlang is a unit of measurement equivalent to one circuit fully occupied for a given period.

Grade of Service

Traffic congestion measurements indicate the number of customers in the system at any particular time. A congestion measurement determines whether customers are being given the proper service. The element of traffic engineering that measures congestion is the grade of service provided to the users. Grade-of-service criteria is given the symbol P (after the French mathematician Poisson, but usually read as Probability).

$P(0.01)$ = Probability that one call in a hundred will be blocked.

$P(>3 \text{ s}) 1.5\%$ = Probability that 1½ percent of the calls will be delayed more than 3 s.

$P(>0)0.02$ = Probability that two calls in 100 will encounter any delay.

Grade of service relates to call congestion within various system stages measured during the busiest hour of the day. The two principal types of call congestion are lost calls and delayed calls. Delayed calls are those waiting for an available service circuit and are normally queued within the switching system, similar to a checkout in a supermarket. Lost calls occur when the called line is busy or a link to the called line is busy or cannot be reached, and the call is not returned to the system for further action.

Busy Hour

Traffic engineering is concerned with the grade of service that users are receiving during the busiest hour of the day—the worst case. Hence, a summer resort would only implement a traffic study during the summer, whereas a department store would conduct a study for the lead up to Christmas. The result of a traffic study provides a typical distribution of traffic and this serves as a key input to subsequent planning.

Traffic Model Formulae

There are several established traffic formulae that can be used in a tele-communications system to predict performance. To use them, a service objective is first assigned to voice calls routed over particular network segment—some measure of calls lost if congestion is encountered. With this service objective and the load determined by the traffic study, service criteria can be derived from formulae developed over the years by Erlang, Poisson and others.

Erlang B is the formula most widely used in Europe for estimating the required number of trunks, circuits, or other devices. Over the years, Erlang B has successfully predicted the actual devices needed for a partic-ular grade of service. The formula assumes random call attempts, a small blocking probability and that blocked calls clear down. It is well suited to large-scale telecommunication networks. The Erlang B formula derives blocking probability on n circuits which has traffic (in Erlangs) of A.

$$P = A^n / n! \text{ divided by the sum from } x = 0 \text{ to } x = n \text{ of } A^x / x!$$

Typically, network planners will use ready calculated tables that give the traffic intensity that can be accepted for a given Grade of Service and number of circuits.

In the USA, the Poisson formula is more frequently used. This formula is slightly more pessimistic than the Erlang formula because, for instance, it assumes that blocked calls stay in the system for some time. There are other formulae that make other assumptions (such as Neal-Wilkinson, which assumes non-random traffic). The results and usage are much the same.

Packet networks

In packet networks, as we have already seen, it is delay rather than block-ing that determine the grade of service. Hence, the traffic engineering is concerned with packet latency, throughout and packet loss.

More specifically, the Quality of Service parameters that have to be supported in any packet network and which need to be signalled in a call setup message include:

- Cell Loss Ratio (CLR)—a measure of how much traffic has been lost compared with the total amount transmitted. The loss can be attributed to any cell/packet-corrupting event like congestion or line encoding errors.

- Cell Transfer Delay (CTD)—a measure of the time required for the cell to cross certain points in the packet network. The primary concern to end-users is the time required for the last bit of the cell to leave the transmitter until the first bit arrives at the receiver. CTD can be effected by processing in the network switches or routers (or, more likely, buffering in the switch/router).

- Cell Delay Variation (CDV)—a measure of how the latency varies from packet-to-packet as they cross the network. Variation is caused by queuing and variation in the routes that packets encounter as they are transmitted. CDV is of concern in packet networks because as the temporal pattern of cells is modified in the network, so is their traffic profile. If the modification is too large, then the traffic has the potential of exceeding the bounds of the traffic profile and some cells may be dropped, at no fault of the transmitter.

Despite being a more recent arrival on the telecommunications scene, the theoretical basis for analysing packet networks is probably more mature than that for circuit switched network. To all intents and purposes, it is a network of queues that is being analysed and there is a wealth of know how in the definitive text by Kleinrock.

Whatever techniques are used, the ultimate aim of all traffic engineering is to determine how best to satisfy the grade of service required by a user population. It is usually carried out by teletraffic specialists who have software packages for configuring communication switches and estimating the number of circuits required in a voice network or routers in a packet network.

4.8 SUMMARY

Hand-crafted telecommunications networks are a thing of the past. It is no longer viable for even the largest of the network operators to develop their own transmission or signalling technology. Standardisation and commoditisation are key aspects in the design and deployment of modern networks.

This implies that you have to understand the key components that are being supplied. Some of these components are virtually universal. For instance, SONET/SDH provides virtually all of bulk transmission, C7 handles all of the inter-network signalling. Other components are more differentiated, such as those on offer for managing networks. Nonetheless, almost all of the technology that is used in telecommunications conforms to some generally accepted pattern.

This chapter builds on the previous one by illustrating how some of the key concepts of telecommunications systems translate into components that can be used to construct them. Needless to say, there is more variety and depth than we could possibly present here. Nonetheless, we have painted enough of a fast moving and diverse area to enable it to be viewed in some perspective.

REFERENCES

Black, Ulyless (1997) *ISDN and SS7—Architectures for Digital Signalling Systems.* Prentice-Hall.

Griffiths, John (1983) *Local Telecommunications.* IEE Telecommunications Series.

Kleinrock, Leonard (1975) *Queuing Systems,* Volumes 1 and 2. John Wiley & Sons.

Manterfield, Richard (1991) *Common Channel Signalling.* IEE Telecommunications Series.

Minoli, Daniel (1991) *Telecommunications Technology Handbook.* Artech House.

Perros, Harry (1995) *Queuing Networks with Blocking.* Oxford University Press.

Schwartz, Mischa (1999) *Telecommunications Network.* Addison Wesley.

Walke, Bernhard (1999) *Mobile Radio Networks.* John Wiley & Sons.

Ward, Ellen (1999) *World-Class Telecommunications Service Management.* Artech House.

5

Key Computing Concepts

There is no life today without software .... The world would probably just collapse

James Gleick

Although little more than 30 years old, the variety of computing and information technology is vast. No-one could grasp all of it in any sort of detail. Yet there are a few key concepts that will sustain you through the most challenging of technical debates. The aim of this chapter is to build a broad appreciation of those key computing concepts that underpin current communication networks and systems.

So, in this chapter we are going to explain exactly what is meant by the terms object orientation, client/server and data warehousing (amongst others). Many practitioners might see these as trendy terms that are over-blown and complicated beyond many people's ken or practical use. But each embodies some enduring or sound way of constructing networked computer systems, so they are worth spending some effort to understand.

There are plenty of detailed guides to each of the topics explained here and most of these run to several hundred pages. Before diving into such detail, it is worth knowing what is there and what it can do for you. In the words of Sir Isaac Newton 'if we see far, it is only because we stand on giants' shoulders'. Despite being a mere youth in engineering terms, there have been quite a few computing giants and their legacy is there for all to use.

We now move on to some of the more prominent ideas in computing that have taken hold and proved their worth.

5.1 OBJECT ORIENTATION

Communications systems are complex entities, comprised of many components. Some way of breaking down and controlling this complexity is essential. And this is where object orientation fits. It provides a systematic way of describing and building complex (software-rich) systems from simpler components, or objects (Figure 5.1).

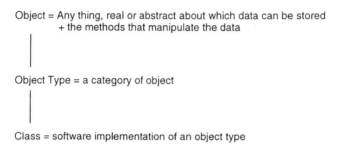

Object = Any thing, real or abstract about which data can be stored
+ the methods that manipulate the data

Object Type = a category of object

Class = software implementation of an object type

Figure 5.1   A basic object

Object orientation has been around since the early 1980s and was originally conceived as a programming approach. In the intervening years the term has become somewhat overloaded as the ideas have been used more widely. To avoid any historical confusion, we will explain the basic ideas of object orientation before going into its application to network and service management.

Objects themselves are no more than pieces of closely related processing and data (something tangible like a modem or something abstract, such as an intelligent agent). To count as an object there must be a well-defined boundary or interface that provides a set of methods for interacting with the object.

There is more to object orientation than just interfaces, though. They provide a basis for describing network components in a common framework such that their interaction with each other can be understood. The basis of object orientation lies in three simple concepts—encapsulation, inheritance and polymorphism.

- Encapsulation   This is all about hiding an object's internal implementation from outside inspection or meddling. To support encapsulation you need two things. First, a means to define the object's interface to the outside world. Secondly, you have to have some form of protocol or mechanism for using the interface. An object's interface says 'this is what I guarantee to do, but how I do it is my own business'. It is useful to regard of the interface as a sort of contract—the object promises to provide the specified services and a user promises to abide by the defined interface. The mechanism for using the interface might vary. It

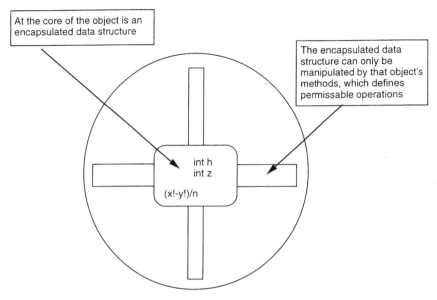

At the core of the object is an encapsulated data structure

The encapsulated data structure can only be manipulated by that object's methods, which defines permissable operations

int h
int z
(x!-y!)/n

Figure 5.2 Encapsulation

could be sending a message or something else entirely—it does not really matter as long as the relevant interface conditions are created. The most widely used mechanisms for communicating between objects are outlined in Figure 5.2.

There are two key consequences of encapsulation. The first is one of simplifying the job of users of an object by hiding things that they do not need to know. The second, and more important benefit, is that it is much easier to maintain a degree of consistency and control by prohibiting external users from interfering with processing or data to which they should have no access.

- Inheritance   This comes in two closely related but subtly different forms. The first form is 'inheritance of definition'. One of the strengths of object orientation is the ability to say something like 'object B is of the same type as object A but with the following modifications'. If you are at all familiar with object-orientated terminology you will probably recognise this as 'object B's class is derived from object A's class'. However it is phrased, the basic idea is that an object's interface can be defined in terms of behaviour inherited from other classes of object plus some modifications, additions or deletions of its own (Figure 5.3).

The advantages of inheritance of definition are twofold. First, there is obvious economy in expressing an interface in terms of a set of already defined portions with a small amount of extra, specific detail. The second advantage is in the lessened potential for introducing errors. There is a caveat on this—you must know exactly what it is that you are

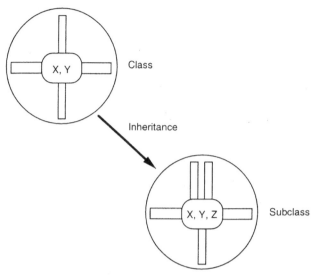

Figure 5.3 Inheritance

building on to avoid inheriting some generic defect! It is easier and safer to specify a small set of changes to an already defined and probably working interface than to respecify the whole thing.

The second form of inheritance flows naturally from the first. It can be described as 'inheritance of implementation'. Suppose you have an implementation of an object of class A and then define a B class as inheriting all of A's behaviour but just adding one extra facility. It would be very useful if you only had to specify the new function and then the B-class re-used all of the remaining A-class facilities automatically. This is precisely what happens in object-orientated programming environments.

In a distributed environment it may be useful to specify interfaces by using inheritance. However, if an object is at the other end of a network there is no way to tell if its implementation uses inheritance or not. Encapsulation applies here as much as anywhere else!

- Polymorphism   The third main object-orientated characteristic is polymorphism. This simply means many forms, and it refers to the ability of an object of one class to be treated as if it were of another class. In object-orientated programming this facility is usually only available to classes related by inheritance. So, objects of class B derived from class A may be able to substitute for objects of class A if they preserve the interface of class A as a subset of their own. It is not usually the case, however, that another class, say C, which has precisely the same interface as class A can be used in its place. This is because the mechanism used to implement polymorphism is usually dependent on the shared code base that comes through inheritance of implementation.

However, things are a little different if we assume a networked situation. If the object at the other end of the wire claims to support particular interface and implements all of the necessary protocols it is impossible to tell if it is a real object of that type or not. It really is a case of a rose by any other name would smell as sweet. So in this environment polymorphism is probably more usefully thought of as compatibility.

Object-orientated technology has been in practical use in the telecommunications network for many years and with great success. Instead of one switch controlling the entire network, the network comprises many individual elements, or 'objects', that act independently. Messages relay instructions back and forth among these elements telling them how to behave for each call. In ISDN, for example, the messages between network elements are carried over a dedicated digital facility called Signalling System No. 7 (C7).

The behaviours that result from this messaging include set-up, call proceeding, alerting, connect, release, and disconnect. Each network element responds appropriately to these messages and performs actions based on a set of embedded switching 'methods'.

The network's object orientation permits the carriers—both local exchange and Interexchange—easily to expand the geographical reach of their networks, upgrade the software of each network element, and enhance customer services without affecting the performance of the entire switching fabric and without having to take the network off-line to isolate and correct problems.

In the networking environment, the collection of object definitions with which a given management system can work is called the Management Information Base (MIB).

The MIB is a repository of information necessary to manage the various network devices. The MIB contains a description of SNMP-compliant objects on the network and the kind of management information they provide. The objects can be hardware, software, or logical associations, such as a connection or virtual circuit. An object's attributes might include such things as the number of packets sent, routing table entries and protocol-specific variables for the routing of data packets on the Internet.

- Attributes   An object is defined by properties called attributes. In the case of the network elements mentioned previously, the attributes include such things as the object's condition (on or off) and the kind of notifications it emits, such as alarms and performance statistics. Printer attributes are usually simpler, consisting of off-line or out-of-paper notifications.

- Class   In the object orientation, the definition of object types is separate from the actual objects. The definition (or class) is a template for the object, determining what it looks like, how it behaves, and how it is accessed. Each object is created from a class and is called an instance of that class. The process of creating an object is called instantiation. A new class can be created from an existing class. Objects that are formed from the new class are essentially the same as those created from the original, having the same data and behaving in the same way, except for any difference specified in the new class definition.

- Class Library   A collection of reusable software components is called a class library. Class libraries make it possible for developers to construct object-orientated applications by picking and choosing what they need.

- Methods   While objects are the things that are manipulated, methods describe what is to be done with the objects. Objects can be handled in a number of different ways. With GUIs, for example, an icon that is associated with a set of files can be grabbed by pointing to it with the mouse, dragged to another icon of a trash can, and dropped into the trash can. This is the way Macintosh users erase their files. In Norton Desktop for Windows, the same function is accomplished by the icon of a shredder. Alternatively, dragging the file folder icon to the printer icon and dropping it there causes the contents of the file folder to be printed.

  When applying this concept to a network management system's GUI, a bridge/router icon displayed on the network map can be grabbed, dragged and dropped to a printer icon. Once dropped into place, the performance information collected by that bridge/router is printed.

  The specific type of performance data that bridge/router is capable of collecting and printing is defined in the object by its attributes. In the future, as object-orientated technology is increasingly applied to network management, a token-ring connection, for example, can be added to an existing bridge/router merely by using the mouse to grab the token-ring network adapter icon from the configuration template and dragging it to the bridge/router icon. By clicking on the token-ring adapter icon, the network manager can bring up that particular object's attributes, which can be modified when necessary. The same procedure would be used to configure the network interface in other ways, even if the network adapter is part of an IBM/SNA computer gateway or a communications controller.

- Messaging   Objects are requested to perform their functions through messaging. On a LAN, for example, the client object sends a message to the server object. The message consists of the identity of the server object, the name of the function to be performed and, in some cases, optional parameters such as passwords and other privileges. The functions are limited to those defined for that object (Figure 5.4).

There are many tools now available to support object-orientated design and development. In addition, as illustrated above, many of the compo-

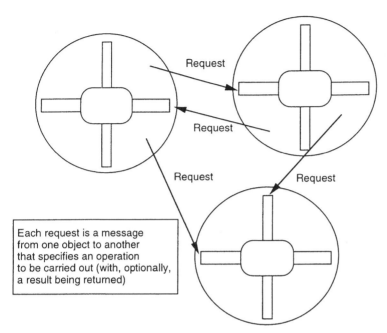

Figure 5.4   Messaging and Objects

nents of modern networks are constructed using OO as a basic principle. Hence this concept is central to virtually all communication systems and networks.

5.2   CLIENT/SERVER

This is, perhaps, one of the most used and abused terms in the computing industry today. It is more than likely to be interpreted by one person in one way, by another in a completely different way. Given this degree of subjective understanding, it is well worth examining the reason for such division before going on to suggest a consensus definition.

First, there is the 'physical client-server' view (Figure 5.5). Here, there

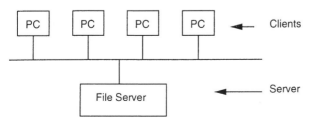

Figure 5.5   Clients and Servers—the physical view

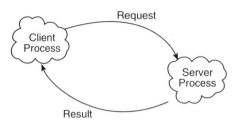

Figure 5.6   Client and Servers—the logical view

are identifiable boxes which are clients (almost certainly PCs) and other identifiable boxes that are servers (probably large Unix machines or mainframes). The machines have the words 'client' and 'server' stamped on their sides, at least in the minds of the IT managers, and it is assumed that the role of client or server is inherent in the type of machine they are.

A second way of looking at the same definition is the 'logical client-server' view (Figure 5.6). Here, instead of thinking about hardware components we are thinking about software components (or objects if you prefer) that exhibit either client or server behaviour. In this instance a particular software component acts as a client and requests services from another software component that acts as a server. There is nothing in the model to indicate where the software components reside physically: they might be on the same machine or different machines of any size or shape. In practice, the server component could be in a PC-shaped box and the client could be in a mainframe-shaped box or vice-versa.

Finally, there is the 'role-based client' server view. Here, a software component might at one time be acting in the role of a client and at other times might act in the role of a server. For instance one object could be a client of one object and a server for another. Hence it could be acting as both a client and a server simultaneously.

So which of these three distinct models is the 'correct' interpretation of client-server computing? The answer is that none of them can claim to be the sole and definitive meaning of client-server. In support of this view we can say that: there is ample precedence for all three usages in computing literature and that there is no single dictionary of computing terms which is agreed to be definitive across the industry and which gives a single definition. If anything, the last of our three is probably the original meaning but it has certainly degenerated through usage to the point where it is pointless to try to reassert any sort of purity of meaning!

In most cases, the meaning of client-server will be determined by the background and mindset of the people using the term and their whole perception of computing. Of course, problems arise when a mixed group of people are placed in the same room and begin discussing the strengths and weaknesses of client-server computing. With people increasingly thrust together by merger, acquisitions and alliances, it is the organisations these people serve that suffer when cooperation is hampered by mutual misunderstanding.

As a working definition, it should be taken that within any given networked computing interaction, one party will initiate the interaction and the other will respond. The one making the request adopts a client role and the one responding that of a server. These definitions are a reasonable starting point, but they do not really paint the whole picture.

Another way of viewing the term server is as a component of a distributed system that provides a re-usable set of services to the rest of that system. The whole point of distributed computing is to have different parts of an application running in different locations. The decisions on what runs where are seldom performed automatically and never on a machine instruction-by-instruction basis.

The most basic principles of modular design lead to an application being broken up into reasonably coherent chunks, each of which provides a set of related services. In a non-distributed system, these chunks might well end up as libraries of re-reusable code. In a distributed system, they will probably end up as independently executing servers. The relation between the server as a supplier of services to the system and clients then becomes fairly obvious as the client is a consumer of those services.

There is still more to the concepts of client and server than we have described so far (hence the frequent misinterpretation). A server is often seen as a passive entity, waiting to answer requests made of it. A client is, by contrast, often seen as the active party, making requests of servers. It is also usually, but not necessarily, the case that an interaction between client and server is synchronous—a client makes a request and then waits for a response.

It is easy to see how the common concept of a file-server on a LAN fits with what we have described here. When people talk about a file-server they usually mean a machine that holds a central store of files which it makes available over a network to be shared by a number of other machines. Each PC or workstation has the illusion that remote files are part of the local file system. This is, of course, a distributed computing application containing client and server components. This is also a good example of how the term server can become stretched and overloaded. It is often used to mean either the software or the machine on which it is running. This is not usually a problem until you start talking about the server (program) running on the server (machine)!

5.3   THE THREE-TIER ARCHITECTURE

This is a network layering concept that starts from the premise that there are basically three types of logic within any application. These three categories are presentation logic, application logic and data management logic.

- Presentation logic   This is concerned primarily with the display of information and interacting with an end-user. Presentation often employs a significant amount of processing—particularly with the availability of ever-increasing power on the desktop.

    When the only means of presenting information was on an 80 character by 24 line display the potential for creative presentation was severely limited. Presentation logic in this environment is often relatively trivial with menus and forms being the norm.

    Today, many users want to interact with applications by point and shoot methods. Busy executives expect to see sales figures presented as glitzy graphs or eye-catching animations. The advent of multi-media facilities will enrich the presentation possibilities yet further. But however impressive it looks, it is all presentation logic.

- Application logic   With so much such effort being put into producing slick and seductive user interfaces it is sometimes difficult to remember that presentation is only part of the story. The real meat of an application is generally independent of the look and feel presented to a user.

    Consider, for example, a typical spreadsheet application. The underlying logic is concerned with applying various functions to tables of data to produce sets of results. Most spreadsheets allow those results to be displayed and presented in various ways— tables, graphs, pie charts and so on. In all cases the supporting application logic is the same.

- Data management logic   If application logic is all about manipulating data in different and interesting ways, there is usually a much more fundamental task to do with storing and managing the basic data itself. Sometimes this is a simple as just being able to read and write data to long-term file storage. In other cases this can be a much more complex task.

    Let us return again to the example of a spreadsheet application. Many such applications have the capability to draw data from tables from a separate database. A spreadsheet might well be used to retrieve a set of sales figures, for example. It could process them in different ways— perhaps calculating monthly averages by sales region—and then store the results back in the database.

    In a multi-user environment, however, many other people could be performing similar or different operations on the same data. The data management logic within the database has to ensure the overall integrity of the data in the face of potentially conflicting requests, perhaps by using locking mechanisms. Another example of data management logic might be the need to ensure that items of data were of the correct format or facilities to filter out data which a particular user was not authorised to see.

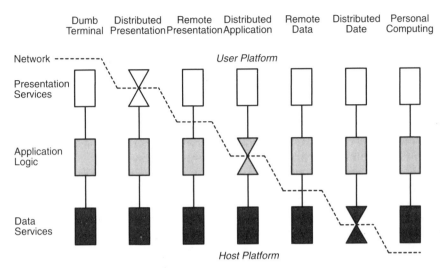

Figure 5.7   The three-tier architecture

The reason the model is described as tiered is that usually presentation logic sits on top of application logic, which in turn draws on data management logic. Presentation is obviously the closest part of an application to a real user and the raw data often seems furthest away. Having identified the three tiers there are obviously several ways in which the function of those tiers can be split between a client and a server. This is shown below in Figure 5.7.

The Figure illustrates where the break between client and server can be inserted for various computing configurations. The portion above the dashed line—which, in reality is the network connection that links the two halves—is logic in the client and below the line is in the server. The 'thickness' of the client increases as more and more logic is placed in the client and less in the server.

The thinnest client, shown at the extreme left, has no logic at all. This best represents the classic dumb terminal attached to a central machine. All the logic resides in the central machine and it is stretching the definition to its limit to call this a client-server system at all, although a 'network computer' would fit in this category. The client becomes progressively fatter until the other extreme is reached where the entire application resides in the client—the typical standalone PC.

Each of these ways of cutting the cake has its own advantages and disadvantages. A thicker client requires more local processing power but less network traffic and less sharing of potentially common code and data. A thin client can be lean and mean but might generate more network traffic and a greater load on a server. Just how these trade-offs are handled is something that a designer should pay attention to as it can significantly affect the performance of the system.

Of course, the next logical progression is to ask the question: why just stop at one cut between client and server? You may already have noticed that each of the splits described above seems to turn the three-tier model into a two-tiered implementation.

Many of the current client-server technologies in the market support this sort of approach. The most common products, such as those using Microsoft's Open Database Connectivity (ODBC), make the split at the data tier. Here fat clients containing large amount of presentation and business logic cluster around database servers. This is what is known as a two-tier or sometimes first generation client-server system.

However, one of the main advantages in recognising the three kinds of logic within an application is that you can start defining boundaries between them. Once you do this you can define interfaces to components which can be reused by other applications. If you lock large amounts of logic into a client it is much more difficult to reuse it for other purposes. For this and other reasons many people favour splitting the three tiers of an application between more than just a single client-server pairing. However, this is where a subtle confusion starts to creep in.

Many people have observed that the old-fashioned mainframe system is very good at some things, in particular managing very large volumes of data. Similarly, personal computers are generally well equipped to perform demanding presentation tasks.

Finally, there has often been an investment in quite high-powered but commodity local application or file servers, perhaps running some flavour of UNIX. Following this line of thought often leads to the following equations, particularly in organisations which already have a large investment in existing infrastructure:

1. presentation logic   = personal computer
2. application logic   = local server
3. data management logic = central mainframe

In many ways this is not a bad first approximation for a wide range of applications. However, it should not be taken as a hard-and-fast rule that one tier equals one box. It is certainly a good design principle to try to separate the different processing rules as much as possible. So, for example, software on a desktop machine might primarily be a client concerned with presentation while a server process may implement a particular set of application logic.

This one-to-one mapping principle is by no means an inviolable rule, though. There are sometimes good reasons for the borders between roles to be a little diffuse. This is particularly the case where performance considerations are paramount. For example, it may be better to place some elements of application logic alongside presentation code in order to avoid excessive client-server interactions. However, the cost of this blurring is that function may not easily be reusable and duplication of effort can result (Figure 5.8).

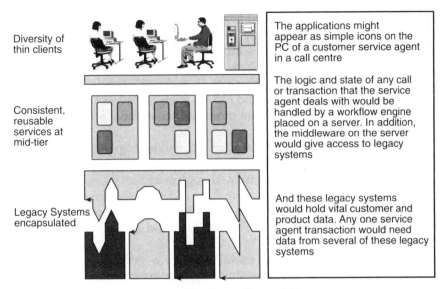

Diversity of thin clients

The applications might appear as simple icons on the PC of a customer service agent in a call centre

Consistent, reusable services at mid-tier

The logic and state of any call or transaction that the service agent deals with would be handled by a workflow engine placed on a server. In addition, the middleware on the server would give access to legacy systems

Legacy Systems encapsulated

And these legacy systems would hold vital customer and product data. Any one service agent transaction would need data from several of these legacy systems

Figure 5.8   A practical view of the three-tier architecture

In practice, different splits of logic will end up in different physical tiers for different applications. For instance, presentation could be mainly in the end-user workstations, application logic mainly in the mid-range systems and data management logic primarily back in the mainframe. However, for some other purpose the split could be very different.

The three-tier architecture is by no means the only 'map' that the designer has for constructing systems. Over the years, there have been a host of proposals for how to structure complex information networks. In the main, they all share some basic principles—separation of concerns, defined interfaces, functional components etc—but tend to be more prescriptive and domain specific than the three-tier architecture.

In the main, those architecture that have taken hold are supported by one of the major players in the computer or communications business. Of those to choose from, perhaps the best known are the Distributed Computer Environment (DCE, developed through the Open Software Foundation, a large international industry group), Telecommunications Intelligent Network Architecture (TINA, developed by a consortium of major telecom companies) and Open System Computing Architecture developed by Bellcore (now known as Telcordia). More on these later.

5.4   MIDDLEWARE

Many of today's large companies have networked systems that are both complex and heterogeneous. The complexity stems from many causes, but mostly from the move to greater levels of distribution. They are typically heterogeneous because not only are the systems designed with a particular purpose in mind, and the most appropriate technology chosen on the basis of which will do the job most effectively, but also on the fact that technology is progressing at such a rapid rate that this year's model is next year's legacy system. Some would call it variety, some would call it 'best of breed'. The bottom line is that most businesses have a huge range of systems to integrate.

Many companies, if they could afford to, would strip out their existing systems and start again and base the systems on standard technologies. But few, if any, are in such a position (though examples do exist). The only alternative is to find a way to make these different systems (which were typically designed as stand-alone, bespoke systems) appear to be one single logical entity. It is with this aim in mind that middleware was developed.

In the computer industry, middleware is a general term for any program that serves to 'glue together', mediate between, or enhance two separate and usually already existing programs (Figure 5.9). In somewhat glib terms, it is the "/" in client/server.

Perhaps the most common application of middleware is to allow programs written for access to a particular database to access other databases. In particular, the middleware serves to make data held in legacy systems readily accessible.

In practice, middleware is software that enables cross-platform, distributed computing (that is, getting computers from different vendors and running different operating systems to communicate). It can also be viewed as application program interface (API—see later) which shields developers from underlying detail and operating systems, along with a

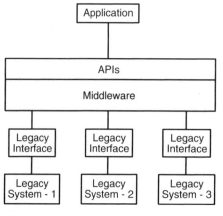

The Middleware provides a consistent interface for the developer to use (a standard set of calls and services from the API). It is connected to various legacy systems, thereby hiding their complexity.

In this picture, the application may be to provide enrol a new user. The developer sees the API with relevent services. In reality, these only work because the middleware can gather the enrolment register from legacy system 1, the user details from 2 and the service details from 3

Figure 5.9   The role of Middleware

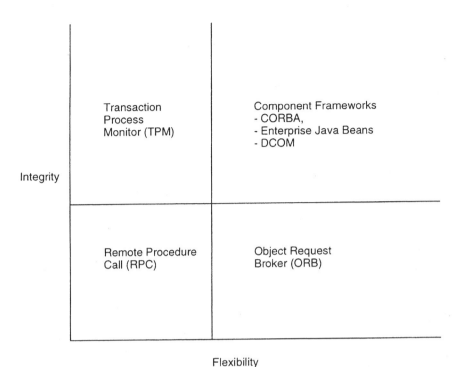

Figure 5.10 A map of middleware technology

mechanism, for example message passing or remote procedure calls (RPC—see later) for cooperating applications to communicate (over a LAN or WAN or within the same machine).

There are various approaches to building middleware, all of which are described in the next chapter. Figure 5.10 positions these technologies in terms of their flexibility and integrity.

The effect is that from a user's point of view an Oracle database looks the same as an IDMS database, and DOS looks the same as Unix. It is this middle layer of software that effectively does all the translation necessary. Hence, it acts as an interface agent.

Clearly, this apparent transparency between systems does not come for free. Middleware can be very complex. What might have been little more than the act of pressing a button on a keyboard to invoke some batch process at a local database, for example, will typically involve a series of programs which in the middleware layer just to reach the 'pressing of the button' stage. The advantage is that all this interface complexity is bundled into a single layer where it can be properly managed and controlled. Later on we will examine the candidate technologies for building Middleware (e.g. CORBA and EJB)

5.5   MOBILE CODE

Most people will have heard of Java. Surprisingly few can explain why it is important from a communications point of view and what are its basic attributes. The idea behind Java, and the more general idea of mobile code, is very simple. It can be thought of as an extension of the World Wide Web concept of 'write once—read anywhere'. With mobile code, the idea is that you write the program once and run it anywhere.

In a sense, there is little difference between the downloading of mobile code and a machine loading word-processor software, for example, from a central file-server on a LAN. However, a conventional word-processor program is distributed in a form compiled only for particular computer hardware, for example, Intel processors. For true mobile code to be achieved then it must be capable of running on any sort of machine.

There are broadly two ways of achieving the required degree of machine independence. The first is to transfer programs round in their source form and to have this interpreted on every machine. This scripting approach works quite well, although it is not necessarily very fast. A major disadvantage is that for large and complex applications the size and slowness of the script becomes too great to make it viable. A second problem is that, if the source of the whole application is being distributed, it makes it very difficult for software authors to retain any ownership and control. Having said this, for small and ad hoc applications scripting is very simple and attractive. Examples of languages that have been used in this way are Tcl and General Magic's TeleScript.

The second approach is to equip different computing platforms with the same virtual machine. This means that a programmer can design for a notional environment that can readily be created on a wide variety of computers. In practice, the 'virtual machine' is a software program which provides an execution engine for primitive instructions from which executable programs can be constructed. The beauty of this is that programs compiled to run on the virtual machine can be distributed, without the need for any changes, and run wherever the virtual machine itself has been implemented. This is precisely what is needed for mobile code.

A developer needs to perform slightly more work and have a few more tools to exploit a virtual machine. Programs have to be compiled from their source into byte-code, meaning an extra step, another tool and some greater configuration management issues. However, the benefits are great. The speed of byte-code interpretation can approach that of native compiled code and there is no need to publish the source code any more—only the byte-code is shipped around and so there is an inherent level of security and copyright protection.

The virtual machine idea is by no means new. In the 1970s the UCSD p-system was a fairly well-known environment used, amongst other things, for making a Pascal language compiler available across a wide range of machines. Smalltalk and other highly interactive languages environments often use byte-code implementations. So why has this caused such a recent stir in the computing world?

In mid-1995 Sun Microsystems released a new World Wide Web browser called HotJava. In many ways this was a fairly standard browser, probably with fewer capabilities than the market leaders. It was written in an object-orientated programming language called Java (originally developed for embedded application such as controlling toasters). Java code was compiled into byte-code and executed by a Java Virtual Machine (JVM). This, of itself, was not what makes the browser unusual. The most exciting feature is the ability of the browser to download Java byte-code from Web pages and execute them. This enables the pages not only to display interesting new forms of information such as animated diagrams but also allows the browser to be extended on the fly.

There are, of course, a large number of potential security issues associated with downloading code of unknown provenance. Not least is the potential for viruses to be imported. The Java virtual machine tackles a number of these issues by placing strict limits on what a program may do. The Java language itself tries to make it impossible to write unsafe programs. The measures adopted address many, although probably not all, of the major problems.

Many people have now seen the potential for the use of Java and a number of companies, including Netscape Communications and Microsoft, have licensed Java technology. The Netscape browser incorporates Java support along with a scripting language, JavaScript. In fairly short order a very large community of users have become Java-enabled (this is a standard feature in current browsers). Downloadable Java programmes (known as applets) are readily available off many web sites.

The two languages in the vanguard of practical mobile code, Java and JavaScript, are linked but independent. Java is an object-orientated language based on C++ that can be used beyond the WWW environment. JavaScript is intended to be a more straightforward language to learn and use and is intended to be used as part of the armoury of the World Wide Web: an entry of <script language = "LiveScript"> in an HTML document indicates the start of JavaScript code, which is then interpreted. The key characteristics of the two languages are summarised in Table 5.1.

Table 5.1

Java	JavaScript
Compiled on server before execution on the client	Interpreted by client
Object-orientated.	Object-based—no classes or inheritance
Applets distinct from HTML	Code embedded in HTML
Strong language typing	Loose language typing
Dynamic binding	Static binding

The range of tools to support Java has grown rapidly over the last couple of years and some of these are explained in the next chapter.

5. 6 DATA WAREHOUSING

This is yet another buzzword within the computer industry. Hundreds of books have been written about data warehousing over the last few years and it is still held up by many as the most significant advance since the database itself.

The idea behind the data warehouse is very simple but, in many ways, quite revolutionary. All the theories of data management that precede data warehousing are driven by a couple of fundamental principles. These are:

- that data redundancy should be eliminated. A raft of entity-relationship and normalisation techniques have been developed to guide the designer in doing this.

- that disk storage space should be minimised. Considerable effort is focused on working out all the ways that people might want to use data and how it could most efficiently be stored to serve this expectation.

In the data warehousing approach, duplication of data is tolerated and efficient storage is not a prime driver. Information is copied from other sources and is made available to end-users, who then use it to support a range of different activities. The complement to the data warehouse is data mining, a means of sorting the information that is available and assembling it into a format that the end-user wants to see.

The key point in the data warehousing/mining approach is that design is no longer concentrated on organising the data but on presenting the data that is available in a variety of different ways.

In keeping with the three-tier principle, the data warehousing environment is made up of three main logical components:

- the acquisition area. This entails the identification of data that is held in other systems and the development of techniques for extracting it.

- the storage area. This is all about specifying databases and/or file formats that hold all of the data to be kept in a way that is easy to access.

- the access area. This includes the variety of access tools that are offered to users who access the storage area.

And, again following the three-tier approach, there are physical aspects that overlie the logical structure. In this instance, the two that matter are the infrastructure (the network, hardware and software that is used to implement the warehouse) and the operations (people, procedures, roles and responsibilities). All of these are considered as part of the data warehouse design.

Once the data warehouse is in place, it can be rendered useful through data mining, which is, in effect, the collation of stored information to answer a user's specific request. There are a range of tools that can be used to carry out data mining, ranging from the simple packages that can be driven by a user without specialised training, through to sophisticated pieces of software that can work out subtle inferences and dependencies. These tools include traditional query managers and report writers, software agents that search for items on the user's behalf, statistics packages such as SAS, data discovery systems such as neural networks, On-line Analytical Processing (OLAP) tools and the search and query tools familiar to many users of the World Wide Web.

5.7 APPLICATION PROGRAM INTERFACES

These are the interfaces by which an application program accesses operating system and other services. An API is usually defined at source code level in the form of defined calling conventions etc. and provides a level of abstraction between the application and the underlying machine. APIs are defined to ensure the portability of the code and to encourage third part software providers to write applications for systems.

APIs are usually built as software that makes the computer's facilities accessible to the application program. All operating systems and network operating systems have APIs. In a networking environment it is essential that various machines APIs are compatible, otherwise programs would be exclusive to the machines in which they reside.

In their early incarnations, APIs were defined at the level that a software engineer would understand. They tended to comprise of set of programming calls that could be made across the interface along with a set of data format definitions. Typical of this genre were the APIs to the UNIX operating system that, if heeded, gave some guarantees of portability across the various flavours of UNIX marketed by different vendors. More recently, API to established platforms such as Microsoft Windows have appeared and there is an increasing drive to define higher level APIs that can be used to interconnect services.

5.8 AGENTS

The concept of intelligent agents originated in the field of distributed artificial intelligence. Research in this area is concerned with understanding and modelling actions in collaborative enterprises. Hence, autonomous intelligent systems, now known as agents, seek to pursue local or global goals, coordinate actions across a group, share knowledge and resolve conflict.

In practical terms, an agent is no more than a piece of software that carries out a particular set of pre-defined tasks. For instance, a mail agent might be installed on a PC to monitor and filter incoming messages or a print agent may be used to queue and direct output according to user preferences.

At a more detailed level the most common architecture that underpins agents is known as BDI—beliefs, desires and intentions. In BDI, an agent's state is modelled by its beliefs. For instance, in a helpdesk these would include information concerning a customer (their name, profile etc.). The agent's desires would be the objectives it is set in terms of completing an appropriate business process (e.g. requesting information from customers) and the desires would be invoked by a combination of context (that is, the current state of the agent) and a triggering event such as a customer query. The agent's intention is the particular goal that it is trying to meet and the plan it has invoked to meet that goal.

As systems increasingly move towards being networkcentric, so the role of agents in managing complexity seem likely to grow. There are five main areas in which agents are used:

- Research, where they can ease the otherwise laborious task of searching databases for particular items of information. A good example of this application is the AltaVista product from Digital, which can search Web pages against certain criteria and report back where matches are found.

- Groupware, where a number of users want to work together in a certain way as a virtual team. The way in which information is shared and feedback delivered can be set up using agents.

- Selling, which is similar to research in that matches to certain criteria are sought on the network. In this instance, the aim is to direct a selected set of information at a third party rather than gather a range of information for one consumer.

- Entertainment, for instance, the use of agents to carry out personal tasks such as selecting favourite articles of a bulletin board or particular sections of an on-line periodical. Fido the shopping doggie is a good example of an agent that is both useful and entertaining.

- Network Management, which is quite a natural thing for agents to do as they are essentially network based and could reasonably be expected to look after their own backyard. Status monitoring and reporting of network components are ideal applications of agent technology.

In addition to these areas of application, there are a number of different types of agent, the main categories being:

- Desktop, including interface agents that help users in configuring their network operating system, their applications and the way in which a set of applications work together.

- Internet, including search services, agents that filter content according to specified criteria, notification services, remote action agents and agents that personalise content to user preference.

- Intranet, particularly automation of processes and support of workgroups but also agents that perform resource allocation in a distributed system.

All of the above are built in much the same way—at a technical level, it would be difficult to distinguish one type of agent from another. This aside, the common link between all agent types and areas of usage is that the rise in connectivity between computers is making them an indispensable management tool.

5.9 SUMMARY

In this chapter we have distilled the essence of the key concepts that pervade computing technology. Some of these, like object orientation and remote procedure calls, are fairly basic concepts for a developer to follow. Some others, such as middleware and three-tier architecture are high-level principles that can be used to guide the computing infrastructure for an entire organisation.

In each case, the aim has been to explain the underlying principles and how they relate to the construction of soundly based computing products. Later on we look at some of the here and now technologies based on these concepts that can be bought, downloaded or implemented. There are, in truth, far too many of these for a complete guide. In any case, many technologies are ephemeral. We have started by focusing on the something more enduring.

REFERENCES

There are a great number of books that provide detail on the topics described above—probably too many to fit in your house (or to be accommodated by your budget). If, however, a particular technical option is just right for the job in hand, here are some recommended texts:

Object Orientation
Szyperski, C. (1998) *Component Software: Beyond Object Oriented Programming*. Addison Wesley Longman.

Client Server
Hart, J. and Rosenberg, B. (1995) *Client/Server Computing for Technical Professionals — concepts and solutions*. Addison Wesley.

Three-tier Architectures
Stallings, W. (1993) *Networking Standards — A guide to OSI, ISDN, LAN and MAN standards*. Addison Wesley, Wokingham.

Middleware
West, S. Norris, M. and Stockman, S. (1997) *Computing Systems for Global Telecommunications*. Chapman & Hall.
Open Applications Group Middleware API Specification.
 http://www.openapplications.org/oamas/loadform.htm.

Mobile Code
One of the many Java/Web books in your local store!

Data Warehousing
Mattison, R. (1997) *Data Warehousing and Data Mining for Telecommunications*. Artech House, Boston.

Application Programming Interfaces
Bal, H. & Grune, R. (1994) *Programming Language Essentials*. Addison Wesley, London.

Agents
Liebowitz, J. and Prerau, D. (1995) *Worldwide Intelligent Systems*. IOS Press.

6

Computing Technology

The last thing one knows in constructing a work is what to put first.

Blaise Pascal

There is probably an even more bewildering array of computing technology than there is of telecommunications technology. This is an industry littered with products that work faster, better or are more open that all of their competitors. And the whole area moves at breakneck speed—it is all too easy to become lost among the trees without ever having been introduced to the wood.

So, if the previous chapter described some of the concepts in the world of computing, this chapter is about putting some flesh on those bones. We will take much the same approach as we did with telecommunications concepts and technology. Rather than trying to catalogue every offering across the full breath of the marketplace, we will choose a few of the key technical building blocks that are used for network-based computing, and explain them in some depth.

The main difficulty in doing this is that there is a significant level of overlap between computing technologies. We share Pascal's problem of knowing what to put first! Notwithstanding, we will start by looking at distributed objects, for no better reason than it is a subject that may well evoke a few memories, having already made an appearance under both the computing and telecommunications concepts headings. Object technology provides a basis for managing networks, as well as a host of computing applications.

Closely related to object technology is the next of our subjects chosen for in-depth analysis, the remote procedure call, or RPC. We have already seen a lot about protocols and how they are used in telecommunications networks. The RPC provides a universal mechanism for distributing computing applications. It takes the distance out of computing, in much the same way that an effective protocol takes the distance out of communications.

The remainder of the chapter covers some of the core pieces of technology that support the effective operation of computer networks: directories, security and transaction processing. In each case, we will abstract from specific implementation details and concentrate on the patterns and principles that should be used in all of the commercial products.

Because we want to give some view of the breadth of relevant computing technology this chapter can only provide an introduction to each of our main candidates. The references at the end will, we hope, provide pointers to further more detailed sources of information.

6.1   DISTRIBUTED OBJECTS

In many ways, objects provide a natural way of describing computer networks. And, as we have already seen, the idea of objects is one that makes a lot of sense in the management of telecommunications systems. The notions of strict interface definition and enforced encapsulation that are fundamental to networks and networked applications are also the bread and butter of object orientation.

Cooperation

Before we go any further we should define what an object is. Computer scientists would define it as something that has characteristics of encapsulation, inheritance and polymorphism. However, in practice an object tends to be whatever it is that your current development tool implements. So within one programming environment there is a very clear notion of what an object is, how it behaves, how you communicate with it and so on and there is a very concrete implementation of these objects within software. Similarly another programming language implicitly defines a similar but different model of what constitutes an object.

Working within a single environment does not pose a problem. Different object models may exist elsewhere but you just ignore them. But when we want to communicate, things are different. A number of different companies and organisations recognised this and set about defining the necessary models and rules for object to cooperate over a network.

And so several initiatives have been kicked off in the search for a common object model. The two that have taken a firm hold across the industry are called CORBA and DCOM. The former is the result of work in an industry consortium, the Object Management Group (OMG) to develop a vendor–independent standards for object interworking. The latter is an extension of Microsoft's proprietary Object Linking and Embedding (OLE) technology into the Component Object Model or COM, the distributed version of which is known as DCOM. More on both of these later.

Indirect objects

All of the object interworking technologies initiatives aimed to solve the problem of different object components cooperating within a single process or a single machine. However, it soon became clear that the same techniques that were needed to solve these problems would open up the prospects of interworking with remote objects.

The fundamental principle adopted by all of the technologies is indirection. Requests between objects are made indirectly through a common intermediary. Interfaces to objects are expressed in a commonly agreed form that allows objects on either side of the intermediary to translate that form to its own internal mechanisms.

Within a single program the intermediary can be as simple as an in-memory table of functions. The target object populates the table with functions that it provides which handle each of the operations on the object and cause the internal mechanism to be invoked. A client object invokes the functions indirectly though their position in the table without knowing or caring about the implementation behind it. Whatever the intermediate mechanism the impact for distributed objects is clear. Once the invocation of an operation on an object is hidden behind an intermediary the operation could just as well be performed remotely as locally.

The most widely used term for the intermediary described above is an Object Request Broker or ORB. This was popularised as a result of the work of the OMG [OMAG-92] which contains a diagram similar to that shown in Figure 6.1.

The main components are the Object Request Broker itself and three different categories of objects.

The ORB

The ORB acts rather like the hardware bus on a computer. It connects all of the objects and mediates communications between them. To be a bit more precise, the ORB provides a logical connection but does not need the same sort of physical implementation that you need for a hardware bus.

The ORB may act rather like a stockbroker and mediate all requests between parties. Alternatively it may act more as a marriage broker, handling the initial introductions but then leaving the parties to get on with it!

Application objects

These are the objects built by the users of an ORB. The ORB and all its associated services and facilities exist so that application objects can be

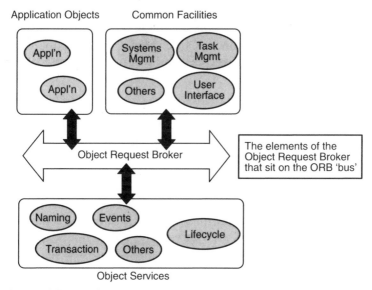

Application Objects   Common Facilities

Figure 6.1   An Object Request Broker

built. However, because the OMG's purpose was to define a common object infrastructure it had little to say about them other than explicitly to acknowledge their existence.

Object services
Object services are fundamental building-block services needed by virtually all applications and which should form part of every ORB implementation. These include facilities such as naming, event notification, transactions, security and persistence (the ability to save the state of an object on to long-term storage in such a way that the object can be reincarnated later on).

Many of these services are not new but one of the key features of the model is that all of the object services should themselves be packaged as objects just like any other. Much of the work of defining object services has been in figuring out how to wrap existing services so that they fit cleanly into the object world.

Common facilities
The common facilities are also collections of objects that provide general reusable services. However, these are facilities that are not necessarily applicable to all applications. They may be thought of as optional packages which supplement the core ORB services.

Some common facilities are intended to be independent of any particular application area. These include user interface and systems management and are known as 'horizontal' facilities, having broad applicability.

Others are 'vertical' facilities applicable to a particular application niche such as finance or telecommunications.

CORBA

The original OMG object model is good for talking about object systems at a high level. It does not, however, give any clue as to how such a system would really be built. Most critically it does not specify any way of defining the interfaces and properties of objects. So it was that in late 1991 the OMG published the Common Object Request Broker Architecture (CORBA) specification [CORBA-91]. The term CORBA is often used as shorthand for much of the OMG's work and indeed it is central to it.

The CORBA specification defines an interface definition with which to specify objects. This language is closely based on C++ syntax but is purely declarative. This means that while you can describe the syntax of an interface you cannot write a program in the language itself. It provides full support for key object-orientated feature such as inheritance. CORBA also describes the structure and principal components of an ORB, as illustrated in Figure 6.2.

Before describing the elements, we should highlight an important difference between ORBs and other client-server technologies. The difference is one of granularity. In conventional client-server technologies there is an implicit assumption that the services called on by a client will be provided by a relatively small number of fairly heavyweight servers. There may facilities such as dynamic startup and shutdown of servers to balance load but there are still reasonably few server instances in existence at any one time. In this world a company's database of customers, for example, may be accessed through two or three main servers. The database may contain tens or hundreds of thousands of customer records but these are all accessed through a few interfaces.

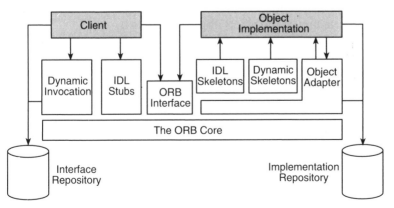

Figure 6.2   CORBA structure

CORBA talks about clients but not servers. Instead it uses the term 'object implementation.' The interfaces defined in CORBA IDL describe a class of objects. A client may need to interact with thousands of objects in the system, all of which conform to the definition of that class. In object-orientated terms, there may be thousands of instances of the object class.

The property of encapsulation means that each object instance is effectively seen as a mini-server, providing access to its unique parcel of functions and data. In the ORB world the customer database described above may be modelled as a collection of customer objects. Each of the traditional database records might be represented by an individual instance of a customer object.

Objects are addressed in CORBA through the use of object references. The structure and content of an object reference is a matter for an individual ORB implementation but it has to contain sufficient information for the ORB to route any requests using it to the place where that object is implemented. As far as a client is concerned every object could be implemented on a different machine. It is the job of the ORB to hide all of this. In practice, many object instances may be handled by a single server program, but you can not assume this to be the case.

Adaptation

To allow the ORB to cope with practical variety the CORBA architecture contains the concept of an object adapter. This forms part of the glue between implementations and the ORB core. Object adapters are responsible for a number of functions including:

- generating and interpreting object references,
- mapping object references to the corresponding implementation,
- automatically activating (and shutting down) implementations when necessary, and
- invoking operations on the object implementations.

Object adapters provide a standard interface to object implementations so that they can perform tasks such as registering themselves as ready to receive calls. The OMG envisage that there will be relatively few types of object adapter. This is wise because portability of object implementations across ORBs depends on having well-defined interfaces to the adapters.

The CORBA specification identifies three likely adapters: the Basic Object Adapter (BOA), a library object adapter and an object-orientated database adapter. The BOA is the adapter that corresponds to starting up a server program or programs to service requests. All CORBA ORBs must support at least a Basic Object Adapter and the interfaces to this are defined in the CORBA specification.

Interoperability

ORB implementations have become available on a number of different computing platforms. Those from a single vendor will happily inter-work amongst themselves but there is no guarantee (and very little hope) of ORB implementations from different vendors working together.

This is where the Unified Network Objects (UNO) specification and Internet Inter-ORB Protocol (IIOP) come into play. The IIOP is a fairly simple TCP/IP-based protocol that defines a number common message formats and exchanges. These allow CORBA IDL data-types, object references and operation invocations to be passed between different ORBs. The message set is independent of the underlying transport mechanism and is also known as the General Inter-ORB Protocol (GIOP). The IIOP provides a touchstone in the CORBA world, to guarantee interoperability with other ORB implementations the IIOP must be supported

COM/DCOM

COM was introduced by Microsoft in 1993 as a natural evolution of the OLE (object linking and embedding) paradigm that they had been using in their suite of Microsoft Office products. Initially intended for application on a single machine using the Windows operating system, it was expanded to allow communication with components on other Windows NT based systems and thus Distributed COM (DCOM) was introduced in 1996.

The Component (or Common) Object Model (COM) and Distributed COM (DCOM) specifications are not standards in the same sense as CORBA, because they are the proprietary offering from a single vendor, Microsoft. However, the ubiquity of Microsoft software on the desktop means that client software in a distributed client-server architecture will almost certainly be a Microsoft application or run on a Microsoft operating system and need to interface with Microsoft applications. Thus in any large-scale enterprise it is very difficult to ignore the need to interface with Microsoft Object Models.

COM 'objects' can be any one of a variety things: a C++ object, an OLE component, a piece of code or a complete application such as one of the Microsoft Office products. The main characteristic of a COM object is its interfaces. There are some basic interfaces which all objects must implement, but beyond that the user can have any number of interfaces. It does not have a component or interface specification language like CORBA, but is a binary standard. Once a COM or DCOM object is complete it is compiled and supplied as a binary executable for the particular environment for which it is intended.

So COM, unlike CORBA is a network interoperability standard. That is, it specifies exactly what code execution mechanism is used to invoke object operations and precisely what the network protocol exchanges are.

This is rather like specifying the details not only of an ORB's interfaces but also its implementation, and, in the context of proprietary standard this is entirely reasonable.

In COM the fundamental interface definition between a client and object is in fact a data structure that sits between the client's code and the object's. The client makes requests to the object indirectly via the data structure. Having said that COM is a binary standard, it does also specify an interface definition language. COM IDL is an enhanced version of DCE IDL; the COM network protocols are also heavily based on DCE RPC (details of IDL and RPC in next section).

COM also differs from CORBA at the level of the basic object model: it has different interpretations for the concepts of class, interface and object. Instead of following the object-orientated model, where an object is an instance of a class, the class is taken to be the mould from which object instances are created.

In COM objects are instances of classes. But, in this case, a class is merely a convenient grouping of a number of related, but distinct, interfaces. So rather than an object being fully defined by its single interface (as is the case with CORBA, where interface and class are virtually the same), it defined by the set of interfaces it supports. Because of this aggregation facility COM makes far less use of inheritance than CORBA. In fact it only supports a limited form of inheritance of interface definitions from a single parent.

It is largely a matter of opinion as to which model is better as the same end results can be achieved with both. If an object needs to provide the facilities of an X interface and a Y interface, in CORBA it will probably define a new XY interface inheriting from two parents. In COM the object will merely support both interfaces directly.

Microsoft has also introduced Active-X controls as components intended for use in Internet applications. Initially these had to be written in C++, but since the release of Visual Basic 5.0 it has been possible to write them in VB. As a result they have in fact been primarily used for Windows desktop application. These two facts have led to a considerable expansion of commercially available components. The model was expanded with COM+ which includes transaction processing services making the components more suitable as a standard for server-side components.

All of these various strands are brought together in Microsoft's Distributed Networking Architecture (DNA) which provides a view of how internet-based applications should operate. DNA covers of business process, storage, user identification and navigation elements and builds on established COM , DCOM and Active-X ideas.

COM and DCOM architectures can only be used on Microsoft platforms so inevitably Microsoft are largely going it alone while most of the rest of the software industry is lining up against them with a converging CORBA and Enterprise Java Beans offering.

Enterprise Java Beans

Java is an object-orientated programming language, based on C++, that compiles into byte-code and is executed by a Java Virtual Machine (JVM). The idea behind Java is that it is 'write once, run anywhere' and the JVM on which the Java runs can be added to virtually any piece of established hardware (including the processor in a domestic toaster). On a more practical level, Java byte-code programs, known as applets, are routinely downloaded into Web browsers for subsequent execution.

Generalising the idea of the applet, we now have components written in Java and these are called Java Beans. Most Java Beans were initially produced as GUI components and did not lend themselves to server-side operation. The rapid increase in the use of WWW front-ends to business systems created an approach based on a thin-client requiring the development of a multi-tier server architecture and non-visual server components. Enterprise Java Beans was therefore launched to extend the component model to support server components.

Visual Java client components operate with an execution environment provided by an application or Web browser. Java server components on the other hand operate in an environment called a 'container'. This is an operating system thread with supporting services that is provided by a major server-side application such as database, web server or transaction processing (TP) system.

A class of EJB is only assigned to one container, which in turn only supports that EJB class. The container has control over the class of EJB and intercepts all calls to it. Effectively it provides a 'wrapper' for the EJB, exposing only a limited application-related set of interfaces to the outside world. The appearance of the EJB to the outside world is called an Enterprise Java Beans Object. The container is not an ORB and client identifies the location of the EJB via the Java Naming and Directory Interface (JNDI). EJB will support CORBA protocols and IIOP. EJB interfaces can be mapped to IDL and can access COM servers.

EJB and CORBA are rapidly converging and it is likely that future versions of CORBA will define the standards for EJB.

UML

Before we move away from the subject of objects, we should make some mention of the key tool of the trade. The Unified Modelling Language (UML) is becoming the natural choice as a modelling notation for component development based on underlying OO distributed technology.

UML is well supported with design and development tools and UML models can be exchanged through the Microsoft Repository. It is not yet clear how it will scale to describe much larger and complex components, how it will support transaction based systems or how well support for process flow will be implemented. However, the extensibility of the notation and the weight of the major vendors behind it almost guarantee its leading role.

UML does not directly address how to address large grained semantic descriptions or how to express non-functional issues. At the moment these tend to be expressed in plain English. These are easy to understand, but also easy to misinterpret and open to ambiguity. For the use of complex commercial components to proliferate such descriptions will need to be tightened up. There are currently no standards or methods defined for doing this. While the extended mark-up language (XML—an extension of HTML) could be a notation for doing this, it is not necessarily the right one and certainly does not address the real issues of how to define those semantics and quality standards. The Open Applications Group (OAG) is working in this area with a view to be able to integrate large-scale applications.

6.2   REMOTE PROCEDURE CALL

Getting computers to communicate could be seen as the straw that breaks the camel's back. After all, it mean having to understand and deal with a whole raft of telecommunications issues—network protocols, blocking, asynchronicity and so on as well an already complex computing world. If you are designing a system to monitor, model and control the processes in a chemical plant then it is a difficult enough problem to begin with—you probably do not really want to have to worry about extending the systems nation-wide and worry about such details as connection-orientated versus connectionless protocols!

One of the most useful general-purpose tools in the computer technology locker is the remote procedure call system. As its name suggests, a remote procedure call, or RPC, makes a remote operation appear just like a local procedure call. Hence applications can be written without reference to the location of resources (assuming, of course, that the distributed object infrastructure is in place).

The remote procedure call is available in just about every programming language in common use. It is the natural way to express a temporary transfer of control 'somewhere else.' It just so happens that in the case of an RPC the 'somewhere else' may be another machine. In terms of the transparencies introduced in Chapter 1, RPC is largely about providing access transparency—a remote access looks precisely the same as a local one.

A good RPC system hides all of the issues of data conversion and network protocols and may also provide other facilities such as recovery from communications failures, location transparency and integrated security. The procedure call illusion is generally maintained on both sides of the remote operation. An RPC client calls what appears to be a local procedure and an RPC server usually implements a matching function that is magically invoked to perform the requested operation. The server procedure returns a result in the normal way which in turn appears as the result of the client's procedure call.

The first practical experiences of building a general purpose RPC system were described in a classic paper by Birrell and Nelson in 1984 [Birrell-Nelson-84]. This paper provided the underlying ideas for practically all RPC systems built since and anticipated nearly every aspect of a complete RPC system design including integrated naming, location and security services. It has only been in the last few years that commercial offerings, such as OSF DCE, have caught up with the initial ideas.

So how is the illusion of a remote procedure call obtained? There are generally three main components to an RPC system: interface definition, stub generation and run-time support.

Interface definitions

For an RPC system the interface definition is the cornerstone of the development process. Most define and use some form of interface definition language (IDL). The purpose of an IDL is twofold. First, it describes precisely what services are provided by servers that support the interface. Secondly it provides the source for generating the 'stub code' which provides the glue between the client and server in an RPC interaction. We will return to the role of the stub code later.

Interface definition languages vary from system to system although many of them share a common heritage and are syntactically very similar. An example of an interface definition using the IDL provided by the Distributed Computing Environment (devised by the Open Software Foundation, now incorporated into the Open Group) is shown in Figure 6.3. The example shows a simplistic interface to a fictitious share dealing service. It provides two operations. The first retrieves a history of the last 10 shares price movements for a specified company. The second function makes a request to trade a number of shares in a given company at a specified price. The actual price obtained is also returned via one of the function parameters.

Although the precise syntax may differ from system to system, the example illustrates a number of features common to many systems. The main components of an IDL definition are generally:

- an interface name for, mainly, human consumption,
- some form of 'globally' unique identifier for system use,
- definitions of the interface procedures and their parameters, and
- definitions of data types which will be manipulated.

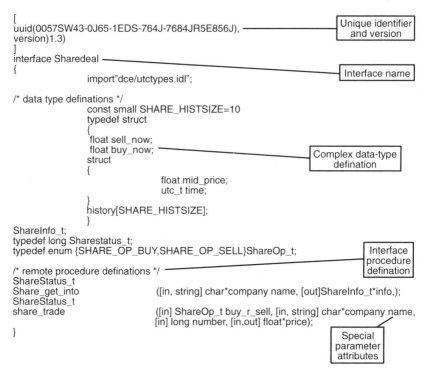

```
[
uuid(0057SW43-0J65-1EDS-764J-7684JR5E856J),
version)1.3)
]
interface Sharedeal
{
            import"dce/utctypes.idl";

/* data type definations */
            const small SHARE_HISTSIZE=10
            typedef struct
            {
             float sell_now;
             float buy_now;
             struct
             {
                        float mid_price;
                        utc_t time;
             }
             history[SHARE_HISTSIZE];
             }
ShareInfo_t;
typedef long Sharestatus_t;
typedef enum {SHARE_OP_BUY,SHARE_OP_SELL}ShareOp_t;

/* remote procedure definations */
ShareStatus_t
Share_get_into                       ([in, string] char*company name, [out]ShareInfo_t*info,);
ShareStatus_t
share_trade                          ([in] ShareOp_t buy_r_sell, [in, string] char*company name,
                                      [in] long number, [in,out] float*price);
}
```

Unique identifier and version

Interface name

Complex data-type defination

Interface procedure defination

Special parameter attributes

Figure 6.3 An IDL specification

The example in the figure shows all of these facilities. Some of these features are probably fairly obvious. You would naturally expect a remote procedure call system to provide a means for specifying what remote procedures are available. Some of the other aspects are more subtle and deal with known problems arising from distribution.

The first problem arises in reliably identifying the interface. If you are going to try to invoke a remote service you would like to have some confidence that you and the remote system will agree on which service you mean. A simple approach is to give the interface a name, such as ShareDeal in the example.

Unfortunately, in any large network where services could be provided by many independent sources, it rapidly becomes very difficult to guarantee that good, mnemonic names will be unique. And ShareDeal is probably the equivalent of Smith in a telecommunications network—if you wish to trade with several stockbrokers each may provide a different interface to their service but all could claim the interface name ShareDeal.

This contention needs to be handled and there are several possible solutions. The first is the registration authority approach. A central body might be set up to own and administer the naming scheme. It could hand out new, guaranteed-unique names on request. The obvious disadvantage of this is that it requires all parties to cooperate with and submit to such an authority. This might prove undesirable or impractical, particularly if there are a large number of transient or essentially private interfaces.

There are, however, circumstances where this scheme works and we have seen a few already—the E164 and X121 numbering schemes for telephone and data networks. Other examples are in network management where interfaces are registered under the auspices of ISO. Similarly blocks of Internet address are assigned centrally by the InterNIC or RIPE (see Chapter 8). In other, non-computing, fields such schemes are in common use for assigning globally-recognised identifiers like International Standard Book Numbers (ISBN) and Universal Product Codes (as shown on bar-code labels).

Another approach to solving the identification problem is to allow human-readable names to conflict but to assign an additional (numeric) identifier to the interface to be used by the RPC system itself to resolve conflicts. The string of hexadecimal digits shown in the example is the DCE solution to this. DCE employs so-called 'Universally Unique Identifiers' or UUIDs which are generated in such a way that it is extremely unlikely that any two UUIDs will ever conflict (or at least, if they do, it will only happen once every million years or so).

There is, however, a further twist to the identification problem. In the good old days you constructed a program by linking your object code with system-provided libraries. This process pulled in the library code and made it part of your final program. The system libraries might subsequently be upgraded or removed but your program would generally continue to run without any dependence on or knowledge of such changes.

Then came shared and dynamically linked libraries. Here the library code is not linked permanently with your program but rather is pulled in as needed at run time. This was hailed as sound practice because it allows program code and memory requirements to shrink. It also allows old programs to benefit from bug fixes or optimisations in new versions of the library. Unfortunately, if incompatible changes are introduced into the library it can also cause existing programs to crash.

A common solution to the shared library problem is to assign a version number to the library. The number can indicate upwards compatible revisions or flag incompatible changes that will break the program. Run-time dynamic linking can select an appropriate version if one is available.

In RPC systems the problem is similar but even more acute. If you own a standalone system you can at least choose whether you will upgrade system libraries. In a distributed environment you may have no control whatever over the system at the other end of the wire. A versioning scheme is also a common solution in RPC systems. The DCE example shows a version number which contains a major and a minor version component. The major and minor versions in this case are 1 and 3 respectively. Under this scheme clients requiring a variant of this interface with the same major version and a lower or equal minor version can interoperate with this server.

The great advantage of versioning interfaces is that you gain a good degree of flexibility in subsequently changing distributed applications. Old clients can continue to operate smoothly to a server using a new, backwards-compatible version of an interface. The clients can be upgraded as necessary in a controlled manner rather than in a big bang approach. Interface identifiers and versioning are very useful tools and represent pretty much the current state of the art in RPC systems. However, perhaps an even more attractive solution in the longer term is the trading mechanism explained later on.

Assuming that you have managed to identify your interface successfully you come to the real meat of the interface definition—the set of operations to perform. In the example, the procedures are defined in a syntax which closely resembles the C programming language. A procedure may return a result (of type ShareStatus_t in the example) and take a number of parameters. Although the RPC procedure definition looks very like a 'normal' procedure definition there are some subtle differences. The most important of these is the addition of some information informing the RPC system how the parameters will be used.

In a local procedure call parameters may be used to provide input to the routine or to provide storage in which results may be received. In particular parameters may point to shared data storage from which the procedure might fetch input data or into which it may place results. However, in an RPC there is no common storage and all data has to be transmitted over the network. So the RPC system has to be told whether a parameter is providing data from a client to a server or vice versa. This is the purpose of the parameter attributes.

The attributes [in], [out] and [in,out] in the example describe whether a parameter is an input value, an output value or both and input and output value respectively. There may be other sorts of attributes which provide further hints to the RPC system such as the [string] attribute which says that the parameter is a C-style null-terminated character string.

The final aspect of the interface definition is the set of data types which can be manipulated by the procedures. Some basic RPC have a fixed repertoire of the commonest data types. In this case an RPC operation is limited to dealing with perhaps a variety of integer and floating point numbers, character strings and so on. This may be sufficient for many purposes. More sophisticated systems allow you to construct and use arbitrarily complex compound types including linked lists and arrays of structures. DCE is such a system and the example shows several examples of this.

Stubs

Interface definitions are useful in their own right as they help to document a set of remote services. However, the real magic in RPC systems comes in the generation and use of so-called stub code.

RPC stubs are the pieces of code on either side of a networked application which provide the illusion of a remote operation happening as a local procedure call. There are two distinct sorts of stub code, one on the client side and the other on the server. The client stub provides a set of locally callable procedures whose signatures match those in the interface definition. When a remote procedure call is made the client stub communicates with the server stub which arranges for the real implementations of those procedures—supplied by the application developer—to be invoked on the server.

Stub code is generated from interface definitions by a tool usually known as a stub or IDL compiler. An RPC client program is created by linking the main body of the application code with the client stub and RPC run-time support libraries. The stub code is treated just like any other module. A server program is created in a similar manner using the server stub code.

In the example given in Figure 6-3, the client stub provides share_get_info() and share_trade() functions to be linked into the client program. However, rather than performing the logic of these operations the functions encode each of their input parameters—those with the [in] attribute—into an appropriate form for transmission across the network. The stubs transmit the data across the network to the server where they are decoded and supplied to the corresponding 'real' procedures as parameters.

When the remote procedure completes the server stub captures the results—the function return value and parameters with the [out] attribute —encodes them and transmits them back across the network to the client. Finally the client stub decodes the data received from the server and the stub function returns exactly as any other local procedure.

Marshalling

The general name for the process of encoding RPC data for transmission is marshalling and the reverse process, converting transmitted data into local data structures, is called unmarshalling.

There are a number of ways in use to represent data transmitted over a network. Hitherto, we have been concerned with issues such as efficiency of line codes. Now the concern is at a higher level, that of the representation of complex structures. These have to be flattened out and broken down into their components. This is often known as serialising : the conversion of a complex structure into a flat stream of bits for transmission or storage. The receiving stub code contains the mechanisms to re-assemble this stream of data into a copy of the original.

Depending on the complexity of the data structures, the marshalling code may also have to manage other issues such as the dynamic allocation and freeing of memory—and this can be a very complex process. However, it is one of the strengths of the mechanism that a good RPC system can even handle the transmission of linked lists, trees and other very complex data structures between client and server without the programmer having to write anything more than the interface definition.

The marshalling processes would end with the disassembly-reassembly mechanism if all machines shared a common data representation. Unfortunately there are wide variations in hardware architecture and so some means has to be found of handling the translation between, for example, ASCII and EBCDIC character sets and different byte orders (i.e. least or most significant bit first).

One approach to solving this translation problem is to have every system converse using a common format before into which everything is transformed before transmission. This approach, used in Sun's External Data Representation (XDR) format, is simple and requires only a single set of encode-decode routines (and, in many cases XDR maps directly onto the native representation used by a large number of machines). Another way of tackling the problem is to adopt a receiver makes it right philosophy, where data is transmitted in native format along with sufficient information for the receiver to be able to perform whatever translation is necessary. This approach, used by DCE Network Data Representation (NDR), has the advantage that if two communicating machines share a common data representation then no conversion is necessary. The disadvantage is that the number of encoding and decoding routines which must be supported is much greater.

Both XDR and NDR encode data largely in terms of primitive data items supported in most programming languages. They can directly represent integers, character strings, double precision floating point values and so on. There is another more generic format called Abstract Syntax Notation One Basic Encoding Rules (ASN.1 BER). ASN.1 is a programming language independent way of describing data structures. The BER describes a way of translating ASN.1 into a data stream which could be transmitted across a network. This format is used in many formal standards particularly those for network management. There is also a draft ISO standard for RPC which uses ASN.1 BER but it is not widely implemented.

Run-time support

RPC stub code can be large and involved, particularly if complex data types are being handled. But producing the code to encode and decode data structure from the network data is obviously only part of the story. The next step is to manage the transmission of the data between the client and the server. This is something that the stub code does in conjunction with the run-time support infrastructure.

Binding

The first and most fundamental task is to establish some form of communication path between client and server. This process is generally known as binding. In a single, non-distributed application a similar binding process often happens when program modules are linked together. It used to be the case that the only option was a form of linking—known as static binding—in which all of the calls between modules were resolved once in building an executable program. All of the modules and pieces of library code were bound together into a single, immutable program.

The analogue of static linking would be to hard-code the network address of servers into clients (rather like having a web browser with all known sites book-marked). This is, of course, possible but the resulting application suffers immediately from a loss of flexibility. If the server needs to be moved to another machine then the client will fail to find it.

In modern systems static linking of programs is frequently replaced by dynamic linking. A dynamically linked program does not contain code from modules or libraries directly but rather has references to modules which are pulled in at run-time. The references often consist of a name, a version and a set of locations in which to look for matching modules. This gives flexibility for the dynamically loaded modules to be changed independently of the main program. They can be modified and moved around provided that the versions remain compatible and the location stays within the known search path.

The ability to bind to a server by name, version and search path is even more applicable in a distributed system. It gives a good degree of flexibility and provides location independence and transparency. There are some additional requirements and opportunities in the distributed case but in many ways dynamic linking and client-server binding are very similar.

One of the complications in a distributed binding is tying down the precise location of a server. Many communications protocols require two pieces of information to identify a channel uniquely. The first is some form of machine address, for example 132.146.116.16 is an Internet protocol address of a single machine. Again this address could be hard-coded but a far better approach is to be able to locate a server by some sort of name that can be translated at run-time to the appropriate address. This is usually achieved by using some sort of directory service (and we say a bit more about directories a little later in this chapter).

Having located the right machine there may be many servers running on that machine each of which is listening at a particular communications end-point (or well-known port in Internet parlance). So how does a client know which server is listening on which end-point?

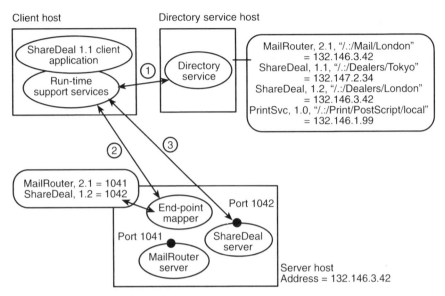

Figure 6.4   The binding process in an RPC system

One approach is to assign well-known end-points to specific servers. This is precisely the approach adopted for common (non-RPC based) Internet services such as telnet and ftp which listen on ports 23 and 21 respectively (hence the epithet 'well-known'). Unfortunately this method is not very flexible or practical except for very commonly used services.

The solution more commonly adopted is to have a special end-point mapping service on each machine. When a new server starts up on a machine it obtains a random, spare end-point and then informs the end-point mapper. When a client wishes to communicate with the server it asks the mapper—which *does* listen at a well-known end-point—for the appropriate end-point for the target.

Figure 6.4 shows how the binding process is carried out in a typical RPC system. The client application needs to use the functions of a ShareDeal server. When the client needs to communicate with a server the RPC run-time code sends a request to a directory service requesting the location of a machine supporting an appropriate server. The directory service finds a matching entry for the server type and returns the protocol type and appropriate network address of its host machine. The run-time support then sends a request to the end-point mapper on the target machine to complete the binding process by finding out to which end-point the desired server is attached.

There are many ways in which the 'right' server can be selected. In the example the client code might specifically have asked for a particular server by name—in this case '/.:/Dealers/London'. Another approach might be for the client to ask for a list of all compatible servers and choose one itself or even just let run-time code handle everything completely transparently. All of these schemes are possible and supported in different ways by different RPC systems.

Communication

Whatever the means of binding, having found the right protocol, machine and end-point a communication channel can be set up between the client and server and the real business of the RPC can finally start.

The basic communications function of the run-time is to take the data buffers created by the stub routines and pass them between client and server. However, in performing this task there are a number of issues the run-time code may have to handle. These include:

- breaking large data buffers up into smaller chunks for reasons of efficiency and to avoid choking clients or servers,

- detecting lost packets and managing time-outs and retransmission,

- handling client or server failure and performing necessary clean-up.

The usual default mode of operation for remote procedure calls is to guarantee something called 'at most once' semantics. This means that if the remote procedure is invoked at all then it will only be executed once for any matching client invocation. The weakness of this is that if an RPC fails part way through there is, in general, no way for a client to know what state the server was in—did the remote procedure execute or not?

The 'at most once' is the usual form of execution as it satisfies most purposes but there are often mechanisms to specify that a particular RPC should have different semantics and which may be more useful under some circumstances.

The most common of the alternate RPC semantic forms are 'idempotent,' 'maybe' and 'broadcast'.

- Idempotent—literally 'same strength'—operations are those that always produce the same result when given the same arguments. A function which accepts an array of floating point values and returns the standard deviation could be declared as idempotent while a function that appends those same values to a file should not. Idempotent operations can be implemented more efficiently than 'at most once' operations.

- A 'maybe' operation is a sort of one-way, no-guarantees call—a client invokes the operation but neither expects a result nor cares if the server fails to receive the call. This may be useful, for example, for notifying interested parties of non-critical status information.

- A 'broadcast' operation usually employs a network broadcast facility to solicit replies from any servers that happen to be listening. It will take the first reply it receives and is mostly only used in boot-strapping operations.

Depending on how the binding process was originally performed it may be possible for the run-time support to rebind to another server in the event of a failure being detected. In any event the run-time is responsible for error detection and for signalling such errors to the calling application in an appropriate manner.

6.3 DIRECTORIES

One of the fundamental problems in any network is how you find things. In a telephone network, you have to know the number of the person you want to speak to, on a data network you need a Uniform Resource Locator (URL) or similar. It is the job of one of the component parts of the system to find services on behalf of a client, and in computing technology RPC systems, ORBs and any others all need this fundamental ability.

Almost every piece of networked computing technology has a location service either explicitly or implicitly provided. The commonest facility is the ability to translate some form of symbolic name into an address. The address need not be a network address but rather is whatever information is needed to tie down the precise location of the service described by the supplied name. The most basic form of this facility is usually known as a name service. You give it a name and it returns its translation, if any. In its simplest form this is little more than a straightforward lookup table.

Finding the directory
Even with the most basic name-service there are several questions that need to be addressed for a networked environment. The first is rather like Russell's paradox: if you use the name-server to find servers then how do you find the name-server?

There are several possible answers to this. One might be to require that clients are supplied with the location of the name-service directly, perhaps through a configuration file, start-up parameter or even by hard-wiring it into application code. This can work quite well although it does suffer from a certain lack of flexibility—if the name-server crashes or has to move the client is stuck until provided with new information. This can, however, be alleviated by providing a list of fall-back servers.

Another solution is for a client to use a network 'broadcast' to find any listening name-servers. This is particularly suited to Local Area Networks (LANs) and provides the ability to find another server in the event of failure. Unfortunately, network-wide broadcast beyond a LAN is usually prohibited, both because of potential delays in getting a response and by the traffic volume it can generate. This makes such a scheme unworkable for wide area use or where any fall-back servers are sited beyond the current LAN. There are also some potentially awkward security issues—if anyone can reply to the broadcast request then they might pretend to be a name-server.

It may also be possible to side-step the problem of locating the name-server. In some environments the name-service is not a separate entity but is completely integrated as part of the supporting infrastructure. If you can talk to the infrastructure at all then you have access to the name-service. The Tuxedo TP monitor's Bulletin Board is an example of this. It exists as a structure in memory and if Tuxedo is running at all then the Bulletin Board will be accessible. If it is not running then clients will not be able to perform any processing anyway, name-service or not.

Accuracy

If you solve the problem of locating your name-service the second main problem arises: how do you make sure that entries are accurate and how you keep the information up to date? If there is a single, central server this is not too much of a problem. It is a reasonably straightforward task to maintain a single database. However, this is then a critical central point of failure. If the name-server is lost the whole system may stop. In reality, you will want the name-service to be resilient to failure. This is harder because you will probably need to replicate the data to other servers.

There are several options for tackling the replication problem— directories are not alone in needing to coordinate distributed information, so the issue has received considerable attention. The easiest solution is to appoint some server as the master and others as slaves. Changes are only made to the master database and then copies are sent out to the slaves when necessary. Sun's Network Information Service (NIS), originally known as Yellow Pages, is of this master-slave type. This works reasonably where there is well-defined central control and management of the whole system. Unfortunately it can suffer from delays in propagating changes out to all slaves. Also if control of the master is subverted then all slaves are subverted too.

A different approach to the master–slave mechanism is to have every name-server remain in contact with every other and propagate all changes to each other. But this is not easy to achieve and veers into the realm of process group technology.

There is a more fundamental issue lurking beneath both the problem of locating the name-service and updating its data. This stems from the move from central control in older computing environment to federation in networked ones. A master–slave or even mutual update strategy can be made to work where there is a single database of information. However, these break down where there is a naming hierarchy with different branches owned by different people.

Probably the largest example of a hierarchical name-service in the computing world is the Internet Domain Name Service (DNS). This is the service used to resolve textual names such as gatekeeper.dec.com and www.inet.org into the Internet addresses 204.123.2.2 and 192.76.6.51, respectively. There is no central DNS server for the Internet. Indeed such a server would be impossible to maintain as it would have to track the changes to potentially billions of network addresses.

As mentioned earlier, the way Internet addressing works is to allocate blocks of addresses and the responsibility for assigning them to particular domains. This is similar to the E164 scheme for telephone numbers and leave Internet Service Providers and large users owning a number of address ranges (128.141.xxx.xxx—where the last two fields are managed locally). All of the addresses in the range correspond to names ending in the suffix '.isis.co.uk' and are part of that domain. The domain names are hierarchical in structure, so the administrator forms part of the .co domain, which in turn is part of the uk domain.

The hierarchical structure is used to direct the lookup process using DNS. A lookup request starts by asking a local server if it can resolve the name directly. It will be able to do this either if it is the authoritative source of the translation for the name or if it has already resolved it recently and has it in a cache. Each server is configured to know the addresses of several other servers and the domains that they handle. If a server cannot resolve the name then it will look at the structure of the name and tell the requester the address of another server more likely to be able to help. This process will continue until either a server with the authoritative data is found or all further possibilities are exhausted.

Directory Services
So far we have been fairly cavalier with the use of the terms name-service and directory. Indeed the terms are often used synonymously but there is a valuable distinction which can sometimes be made. A name-service is something which provides a simple, translation lookup. You specify a single key and return a single record of information. DNS is tailored to performing domain name to address and address to name translations although it can handle other forms of data. NIS can be used for storing and retrieving arbitrary key-value pair information. In neither case, however, can you easily perform sophisticated matching or searching.

Directory services, by contrast, are usually meant to be considerably more flexible. They may allow sets of attributes to be associated with names and provide facilities for performing complicated searches and matches in addition to straight one-for-one lookups. Probably the most comprehensive directory service specification is the ITU-T X.500 standard for electronic mail. This provides a fully hierarchical, federated directory service allowing some very sophisticated operations.

The DCE Cell Directory Service (CDS) is somewhere between a name-service and full-blown directory service. It provides a very comprehensive hierarchical name space with facilities automatically to replicate and synchronise parts of the hierarchy. It can also store arbitrary attributes associated with each entry. CDS does not itself, however, provide full directory searching facilities although libraries layered on top of it add some of these features.

Brokers and traders

Directories and name-services are a necessary part of a network based computing infrastructure but neither is particularly intelligent. They both allow information to be stored and retrieved based on some selection criteria but neither has much knowledge about the meaning of the information that they hold.

The next phase in the development of directory services is to enable them to act rather more intelligently in responding to requests. Described earlier, ORBs are starting to move along this road. In an ORB the role of symbolic name translation is reduced and most resolution operations are of object references. These are opaque structures to the programmer but serve as identifiers for objects known to the ORB. In addition the Interface Repository facilities allow much more dynamic discovery and resolution of services.

Beyond this lies the concept of 'trading'. This idea came to prominence in the ANSA project (see Chapter 9) and from there has gone forward to form part of the ISO Open Distributed Processing (ODP) standards. A trading operation is more than just a lookup of information. Instead it is a more dynamic service that matches offered services with requests from clients.

The first step in trading involves ensuring that interfaces are at least syntactically compatible. This is similar to the facility of DCE and other systems to select a compatible interface based on interface identifiers and version numbers. A trader, however, may have more access to interface definitions and could determine for itself what is syntactically compatible. Trading goes beyond this in also allowing other aspects such as desired quality of service to enter the equation. These can also be negotiated such that a client would like, for example, response times under 0.1 s but could tolerate 0.5 s if no more suitable server was available. The idea of trading is not only to find a compatible service but also the most suitable service.

6.4 SECURITY

The term 'hacker' (in the sense of a malicious, unauthorised user) is one invented by the computing community. When a computing system is networked, there are many more ways in which it can be attacked. Different attackers will have different aims, so there is variety in the forms that an attack on a networked computer system can take. The main ones are:

- Direct Attack—attacker aims to log on to company application and use it as though they were a legitimate user, but for covert purposes. The attack might involve stealing or guessing passwords, using Operating System or Application 'backdoors' or subvert user authentication procedures.

- Denial of Service—prevents a company network or application (e.g. a web server) operating correctly. There are plenty of options to assist the hacker in their quest to deny service to rightful users.

- Loss of privacy—where data is tapped in transit and this is used to damage a reputation or for criminal purposes. Local Area Network sniffers and public network datascopes are readily available and can be used for such illicit purposes.

- Data Modification—data in transit could be modified (e.g. a purchase of £1000 could be made, but the attacker could modify the data to show that only £1 had been spent).

- Masquerade—the attacker simply pretends to be the legitimate host. This could be a web site with a similar URL, designed to defame a company. (The hacker could also use more sophisticated techniques to divert traffic from the legitimate site to the masquerade site.) It could be a host simulator, tricking remote workers into revealing passwords.

- Information gathering—often the prelude to one of the above attacks. Sophisticated scanning tools can be used systematically to search a host for security vulnerabilities—many such tools can be freely downloaded from the Internet!

To counter these threats there are several essential pieces of security technologies. Probably the most fundamental is that of data encryption. Many people have devoted countless years to devising ever more subtle, sophisticated and effective forms of encryption, so it is worth looking at the main ideas that are used.

Encryption
The basic function of encryption is to transform data into a form which hides its original content and meaning while being able to recover the original data when required. It is also usually a requirement for any reasonable encryption system that this reverse translation should only be possible by someone possessing the necessary key to unlock it.

Some form of encryption is the only way to guarantee confidentiality of data transmitted over an unsecured network (and most networks should be assumed not to be secure). The strength of the guarantee depends on the degree to which the original data is concealed by the encryption mechanism and, vitally, the security of the secret keys. The most difficult form of code-breaking is to spot some form of pattern in encrypted data that might allow the coding to be 'reverse-engineered' back to the original data. Sometimes a more productive approach is to attempt to guess a poorly-chosen secret key.

Most encryption algorithms, including the US Data Encryption Standard (DES) and the International Data Encryption Algorithm (IDEA), are called private or symmetric key systems. The security of the encryption depends on a shared secret known only to the two communicating parties.

There is an alternative to private key encryption called public or asymmetric key encryption. In public key systems, such as the Rivest-Shamir-Adleman (RSA) system, a user has a pair of keys—a private key which is secret and a public key which is made widely available. A message encrypted using the public key can only be decrypted using the private key and vice versa. This means that you can receive encrypted messages from anyone who knows your public key, safe in the knowledge that only you can read them. It also means that if you encrypt a message using your private key I can be guaranteed that it could only have come from you if it decrypts properly using your public key.

Public key encryption is a fundamental part of the widely used Pretty Good Privacy (PGP) program. Encryption is usually only applied sparingly because a strong encryption algorithm usually requires many processing steps to implement (and hence time to encode and decode data). It is often used to protect only critical data, such as the distribution of private keys.

Hashes, digests, signatures and fingerprints
There is always a measure of uncertainty in any exchange of data between two parties. How can a recipient know that the data received is the same as the data sent? How can they know if it has been accidentally corrupted or maliciously tampered with while in flight? A common aid to solving this problem is the creation of a sort of digital fingerprint for messages which can easily be verified. This is often known as a hash or digest.

The principle of creating a message hash is simple. All of the data in the message is fed into a procedure. This procedure scrambles the contents down to string of, usually, 128 or 256 bits. This string of bits is the hash value—in effect, the messages fingerprint. The key to a creating a useful message hash it to devise a routine which makes it both very unlikely that two sets of input data will give the same hash value and also extremely hard to construct a set of data that will generate an arbitrary hash value. Even if the input data differs only by a single bit the hash should be different.

Once a hash value has been calculated it can then be attached to a message. When the message is received the hash is recalculated and compared with that supplied with the message. If the communication channel is secure then this will suffice. Where the link is not secure then it is sufficient to protect the hash value by encryption to be able to guarantee the integrity of the message data. An encrypted hash attached to an unencrypted message is sometimes known as a message signature. With public key encryption the hash can act very much like a real signature guaranteeing the origination of a message. The hash is encrypted using the signatory's secret key. If it can be decrypted and verified using their public key then the message can only have been signed by the holder of that matching secret key—the signatory.

There are a number of hash algorithms in widespread use. These include RSA Data Security Inc.'s MD4 and MD5 algorithms (published as Internet RFCs) and the US Federal Information Processing Standard Secure Hash Algorithm (SHA-1).

Authentication systems
Probably the most necessary security service for a distributed system beyond simple encryption is an authentication service. It is the job of such a service to help communicating parties to discover and verify each others' identity.

By far the best-known distributed authentication service is Kerberos, a trusted third-party authentication mechanism. It depends on having a central, highly trusted server to hold the secret keys of all users. It is the job of this Key Distribution Centre (KDC) server to issue so-called tickets which allow users access to all other services. Both clients and servers implicitly trust this server to issue valid tickets. The basic operations of the Kerberos mechanism are illustrated in Figure 6.5.

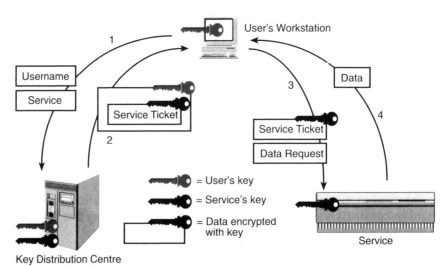

Figure 6.5 Tickets and Keys as a security measure

The process (somewhat simplified from the real thing) is as follows:

1. A user requests a ticket for a service from the KDC.

2. The KDC sends back a packet of data that contains the ticket, but encrypted using the requester's secret key. The only way to unlock the ticket is by knowing the secret key; it will be useless to anyone who does not. This means that a user can obtain a valid ticket without their password having to travel over the network.

3. The user makes a request to the service and presents the ticket it has extracted from the KDC's reply. The ticket itself is a packet of data encrypted using the secret key assigned to the service. This too is known to the KDC that created the ticket. The ticket contains the requester's identity along with other items to guard against forgery and reuse.

4. If the service can decrypt the ticket it will decide whether and what reply to make to the requester.

This is a rather simplified description of the Kerberos protocol (see Schilller's (1994) book for a full exposé]. The most important omission from our description is the fact that the KDC itself only issues a user with a single ticket called the Ticket Granting Ticket. This is then used with another special and highly-trusted service called the Ticket Granting Service (TGS). This exists to circumvent the problem of the users otherwise having to use their secret key every time they need to obtain a new ticket from the KDC. In all cases Kerberos tickets have a limited lifetime to prevent them from being stolen and reused.
Implementations of Kerberos are freely available and are widely adopted and used. It is, for example, at the core of the OSF DCE security services.

6.5 DISTRIBUTED TRANSACTION PROCESSING

In ordinary business terms a transaction usually involves an exchange of some sort, for example of money for goods or services. Examples of transactions that are usually handled by networked computers include making an airline seat reservation, taking an order for a new telephone line, electronic funds transfers and stock control. A feature of all of these transactions is that they have to be carried out with a high degree of reliability—people would be very upset if something went wrong.

Transaction processing (TP) is a specialist area of computing that makes sure that a deal is struck correctly. At the heart of transaction processing are a set of qualities, Atomicity, Consistency, Isolation and Durability, known as the ACID properties. The meaning of each of these properties is described below.

Atomicity

Atomic literally means 'can not be cut or divided'. As far as TP is concerned a transaction is an operation or sequence of operations which either succeed as a whole or fail as a whole.

The usual example of such an atomic operation is that of a credit-debit funds transfer. If I transfer 100 Euro from my account to yours the first step of debiting from my account must be accompanied by the second step of a credit to yours. If the first step succeeds but the second fails I will be 100 Euro out of pocket and you will still be hounding me for payment. If the credit operation is done but the debit fails then we will both be rather happy but a bank as sloppy as this is likely to be heading rapidly for financial disaster.

A transaction processing system ensures that either all the steps in a transaction are completed or that none are. In contrast to the at most once semantics we saw with the RPC it provides exactly once semantics. If all steps complete then a transaction is said to be committed. If any of the steps fail then it is said to be aborted. When a transaction is aborted any partially complete operations are undone or rolled back to leave the state of the system the same as it was before the operation was attempted.

Consistency

The property of consistency is closely related to atomicity. Particularly in the case of failure and roll-back there is a need to ensure that data is left in a consistent state. This usually involves the use of journalling—making a log of the operations performed and the changes caused—at key points. This provides means to return the system to a consistent state if necessary by undoing failed sequences of operations.

Isolation

Transactions are isolated in the sense that the atomicity and consistency measures allow the transaction to execute by itself, without interference from any other work. In practice the transaction may well be executing in parallel with others but the key property is that the effects of one transaction should not be visible to another until the transaction commits. In order to obtain isolation one of the key facilities employed is that of resource locking—preventing other transactions from modifying a resource while a working transaction holds a lock.

Durability

Durability is all about ensuring that once a transaction has committed, the effect of the transaction is permanent and immune from network, system, disk or other failures. Once again, journalling plays a large part in this.

In order to understand how the ACID properties are provided by trans-

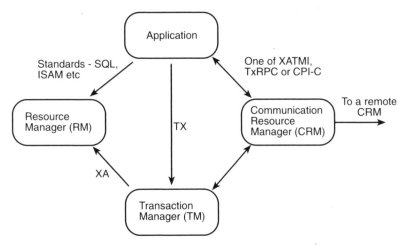

Figure 6.6 The X/Open TP reference model

action processing systems we will start by looking at the X/Open Transaction Processing Reference Model. This model applies both to distributed
and non-distributed applications. The elements of this model are shown
in Figure 6.6. The basic components of the model are as follows:

- Application   The application code is the program provided by a user
  to solve whatever the problem at hand may be. The application draws
  on the services of the Transaction Manager and any number of Resource
  Managers to do its work.

- Transaction Manager (TM)   The Transaction Manager is responsible
  for performing all the necessary coordination necessary to maintain the
  ACID properties of a transaction. When an application wishes to start a
  transaction it informs the TM. It calls the TM again when it wishes to
  end the transaction by committing or aborting. X/Open specify a
  simple programming interface (known as the 'TX' or 'Transaction
  Demarcation' interface) to perform these functions. It consists of
  operations such as *tx_begin()* and *tx_commit()*.

- Resource Manager (RM)   The Resource Manager is responsible for
  maintaining a recoverable resource. A recoverable resource is one
  which can journal its state and perform commit and roll-back operations. The most common RMs are in fact databases such as Sybase or
  Oracle but they could also be other resources such as file-systems or
  print queues.
    Each RM has a native interface used by applications for performing
  operations on its resource. For example the native interface for many
  databases is SQL. The RM also has an interface to the Transaction
  Manager. When the application informs the TM that a transaction is

starting the TM contacts each RM to inform it of the fact. The RM can then associate subsequent actions performed through its native interface with that transaction. Similarly on commit or abort it is the TM which contacts each RM to inform it of the end of the transaction and performs the coordination necessary to maintain the atomicity property. The interface between the TM and RM is specified in the X/Open 'XA' interface.

- Communication Resource Manager (CRM) All of the components described so far are present on a single, non-distributed system. To extend the transaction to another system requires a special Resource Manager called a Communications Resource Manager.

  The CRM is informed of new transactions by the TM like any other resource manager. However, the CRM is also capable of communicating with another, remote CRM to inform it of the start of a transaction which will draw on remote resources. The remote CRM propagates the transaction to its local TM which then informs the local RMs of the new transaction. There are a variety of standard interfaces to CRMs but the interfaces between CRMs are generally proprietary.

The hard work of providing the ACID guarantees falls to the Transaction Manager. If an application uses only a single resource then the facilities provided directly by the resource will be sufficient and there is little need for a TM. For example, a database will provide commit, abort and rollback commands. However, to maintain the ACID properties across updates to multiple RMs, and particularly across multiple systems, requires the use of a so-called 'two-phase commit' protocol.

Two-phase commit
It is not essential to understand how two-phase commit works in order to employ distributed TP technology. However, some appreciation of the underlying process may help in recognising the benefits and limitations of the technology.

To understand how two-phase commit works we first need to introduce the notion of local and global transactions. A local transaction is whatever native unit of work an individual Resource Manager may handle.

When a request is made to the Transaction Manager to start a new transaction it generates a new identifier for a global transaction. This identifier is guaranteed to be unique. The TM then communicates the global identifier to each participating RM so that it can associate subsequent local transactions with this identifier and hence make them part of the global transaction.

To participate in the two-phase commit as well as being able to recognise a request to begin a global transaction a RM also has to be able to handle requests to prepare for commit, commit and abort such transactions.

When an application indicates to the TM that it has finished a transaction by calling *tx_commit()* the TM first informs each RM that the specified global transaction has ended and that subsequent native operations should not be associated with it. If any errors are detected at this stage, such as not being able to contact a RM, all of the other RMs can be asked to roll-back their work. The TM then continues with the first phase of the protocol in which it asks each RM to prepare to commit all the local work associated with the current global transaction. The preparation consists of writing all essential information to stable storage to guarantee the durability property. When the information has been stored it must be able to survive system and communications failure.

The TM collects the responses to the prepare request from each RM. It may be that an RM is unable to guarantee the durability of its information, for example if disk space has been exhausted. In this case it will reject the prepare request. The TM will then ask all the RMs to abort the transaction and roll-back. Similarly if any RM is uncontactable, subject to retry and timeout strategies, the transaction will be aborted.

Only if all the RMs reply successfully to the prepare request will the TM issue the final commit instruction. Once the prepare phase has been passed it is guaranteed that any final request to commit will succeed. If an RM crashes after completing the prepare phase it will be able to contact the TM as part of the recovery process when it restarts to discover whether it should continue committing the transaction or if it should abort it. Similarly the TM will also have logged sufficient information such that if it crashes it will be able to complete the process on restart.

The single biggest problem facing the use of TP in a network is that of locking. In order to preserve the ACID properties of consistency and isolation it is usually necessary to apply some form of exclusive locking mechanism. This protects resources against potentially conflicting, simultaneous updates. A transaction is only allowed to modify a resource when it alone has access to it. Other transactions needing access to the same resource are locked out and have to wait for the transaction currently holding the lock to complete its work and release the lock.

The main issue that arises from locking is the potential to grind a system to a halt. While a resource is locked all other transactions needing access to that resource have to wait. If the lock is held for too long by individual transactions—or worse, accidentally never released—a queue of waiting transactions can build up and response times grow without bounds. There are also the difficulties of avoiding deadlocks where two transactions are each blocked waiting for a lock held by the other to be released.

For all that, TP systems provide a useful piece of the jigsaw in a networked world. To demonstrate the circularity of computing technology, the Encina TP system (from Transarc Corporation, a subsidiary of IBM) is built on top of RPC mechanisms, as explained earlier.

6.6   SUMMARY

This chapter has covered a few of the many computing technologies in current use. Given the sheer scope of the area, we have concentrated on those areas that deal with the distribution of computing services over a network. Starting with distributed objects, we have worked through the operation of the remote procedure call and the way in which directories are implemented. To finish, we considered how security can be assured on a network that is open to attack and how the integrity of transactions is protected when network reliability is less than perfect.

In each case, we have given a fairly detailed exposé of how the technology works. So we have illustrated, for instance, how an RPC is used to build a distributed information service and how secret keys can be used to ensure that information sent over a network is kept secret.

Specific implementations based on the subjects explained here bound. Last year's industry leader inevitably becomes this year's legacy problem. Our approach, therefore, has been to steer clear of the implementation-level detail and to stick to the logical basis for each technology.

REFERENCES

Birrell, A. D. and Nelson, B. J. (1984) Implementing Remote Procedure Calls. *ACM Transactions on Computing Systems*, **2**, Feb. 1, 39–59.

The Common Object Request Broker: Architecture and Specification. Revision 2.0, July 1995, updated July 1996

Elbert, B. and Martyna, B. (1994) *Client/Server Computing: Architecture, Applications and Distributed Systems Management*. Artech House.

Orfali, R., Harkey, D. and Edwards, J. (1996) *The Essential Distributed Objects Survival Guide*. John Wiley & Sons.

Schiller, J. I. (1994) Secure Distributed Computing. *Scientific American*, November, 54–58.

Whyte, W. (1999) *Networked Futures: trends for communication systems development*. John Wiley & Sons.

X/Open and OMG. The Common Object Request Broker : Architecture and Specification, ISBN 1-872630-90-1.

7

The Intelligent Network

It is often safer to be in chains than to be free

Franz Kafka

Long, long ago, in the early days of telephony, there was an operator who dealt with a caller's request for a connection. The operator was only really meant to complete the call. In practice, though, they did a lot more. They could sometimes tell the caller whether the person they were calling was available. If not, they could often redirect the call (or even provide the information the caller was seeking). It was the operator's presence that made the early telephone network an intelligent entity. Since then, technology has tried to recreate the role of the operator. Intelligent networks seek to vest some discretion over call handling within the network.

Whether intelligence should be built into the network is a point of debate. The invention of the automatic switch, by Kansas undertaker Almon B. Strowger, was said to have been triggered by some unwanted intelligence—an operator whose sympathy lay with one of his competitors. For all that, Intelligent Networks are a major feature of most Telco's plans and they do add a new dimension to the capabilities of established networks.

This chapter describes the main components and concepts of an Intelligent Network and illustrates how it operates. In doing this, we build on an idea that has been mentioned several times so far: the separation of signalling and user information. We will see that it is this separation that allows intelligence to be added to the switched network.

There is, inevitably, a significant amount of terminology associated with Intelligent Networks, so our first foray into the subject will explain what it all means. First, though, some rationale for building an IN.

7.1   WHY NETWORKS NEED TO BE INTELLIGENT

In principle the existing telecommunications network can provide customised services. It can, for instance, make part of the public network behave as though it were the company's own private network, that is, a Virtual Private Network. In practice however this flexibility has not been achieved with the likes of the ISDN. The potential flexibility of stored-programcontrol switches has not been realised because of the way the call control software and its associated data has been implemented in the exchanges.

The problem stems from the fact that the service information relating to a customer's lines is stored in the serving Local Exchange. The companies with the greatest needs—those with the most to gain from customised services—are large and spread over many sites, indeed often over a number of countries. So the service information relating to such companies is distributed over a potentially large number of Local Exchanges, often several hundred. Indeed, when looking at the collective requirements of national organisations, information is distributed over just about all Local Exchanges, perhaps thousands. The problem of managing such a large distributed database and the associated coordination of customised call control has proved prohibitive.

The solution to this coordination problem has been to separate the 'advanced' service logic and the associated customer information from the 'basic' call control logic and switches. Basic call control continues to reside in the Local Exchange. But the advanced service logic defining the customer's requirements is centralised in what is termed an 'intelligent' database. Adding this centralised network intelligence to the established network infrastructure creates what has become known as the Intelligent Network, or IN (Figure 7.1). As we will see, with this arrangement it becomes comparatively straightforward to manage a comprehensive, upto-date picture of a corporate customer's 'private' network requirements

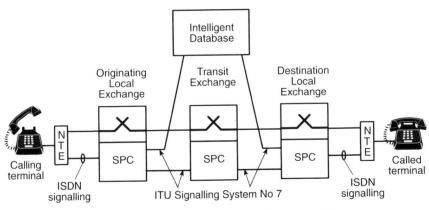

Figure 7.1   Access + Centralised Network Intelligence = IN

and to coordinate switching operations throughout the network in order to implement these requirements. C7 (explained in Chapter 4) provides the signalling system of choice for IN operations.

7.2 IN ARCHITECTURE AND TERMINOLOGY

The main building blocks of the IN are the Service Switching Point (SSP) and the Service Control Point (SCP). The SSP is (usually) part of the Local Exchange whose call control software has been restructured to separate basic call control from the more advanced call control needed for Intelligent Network Services (this terminology is somewhat circular—Intelligent Network Services are simply those services that need the Intelligent Network capability) (Figure 7.2).

Basic call control looks after the switching operations that take place in an exchange. It has been restructured to incorporate what are known as Points In Call (PICs) and Detection Points (DPs) as defined points in the basic call control state machine. At these points certain trigger events may be detected. If this is the case, normal call processing is suspended, at least temporarily, whilst reference is made to the centralised Service Control Point (SCP) to find out how the call should be handled from that point.

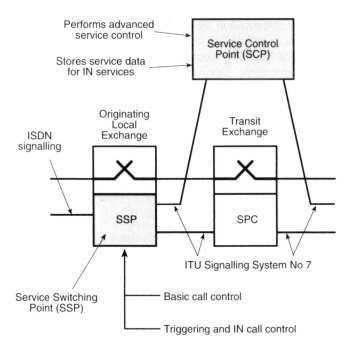

Figure 7.2 IN Architecture and Terminology

Typical trigger events include such things as recognition of the Calling Line Identity (CLI) and recognition of a specific dialled digit string (e.g. an 800 service).

The Service Control Point (SCP) is a general-purpose computing platform on which the advanced service logic needed for Intelligent Network Services is implemented, together with the information that defines each corporate customer's network services. It must be fast to provide the rapid response needed and to handle the potentially very high traffic levels arising from its central location. And of course it has to be reliable. To meet these stringent requirements more than one SCP is normally provided. In practice there may be a dozen or more.

We now introduce a bit more jargon. The Service Switching Point software within the exchange consists of a Call Control Function (CCF) and a Service Switching Function (SSF) (Figure 7.3). The Call Control Function looks after the basic call control needed for simple telephony switching operations. The Service Switching Function provides the control interface with the Service Control Point (and with another IN network element known as the Intelligent Peripheral (IP) that we will look at a little later on). And there is yet more jargon! The Service Control Point (SCP) contains the logic that is needed to implement Intelligent Network Services, as shown in Figure 7.4. Each of these services, such as a Freephone call (which we will look at in more detail below), requires a Service Logic Programme (SLP). This SLP is built from Service Independent Building-blocks (SIBs) together with the service information defining the corporate customer's detailed requirements which is held in the associated Service Data Point (SDP). Service Independent Building-blocks would typically include such operations as number translation, connecting announcements, charging, and so on. Strictly speaking the Service Data Point need not be co-located

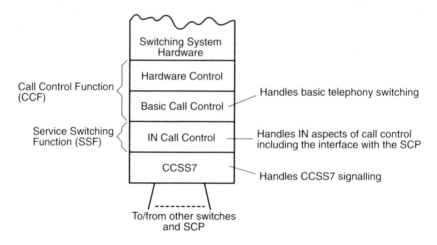

Figure 7.3   The Service Switching Point (SSP)

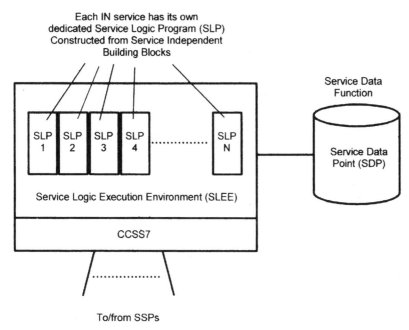

Figure 7.4   The Service Control Point (SCP)

with the Service Control Point. But it usually is and we will assume here that the Service Data Point resides within the Service Control Point.

The Service Logic Execution Environment (SLEE) is the generic software that controls the execution of the Service Logic Programmes. It interworks with the basic call control process and the simple switching functions in the Service Switching Point and screens the Service Logic Programmes from the low-level SCP-SSP interactions and hence it controls the impact of new Service Logic Programmes on existing IN Services.

C7 signalling needed some extension in order to support IN Services (Figure 7.5). In particular a Transaction Capabilities Application Part (TC) was added to allow signalling that is not related to switched connections. At this point, we can complete the picture for ISUP. Since some ISDN supplementary services involve signalling that is not related to switched connections ISUP may also use the services of the Transaction Capabilities Application Part as shown.

Examples of non-connection-related signalling include Operation Administration & Maintenance (OA&M) messages, customer-to-customer data transfer (via the signalling subnet), IN applications such as signalling between the SSP and SCP, and signalling for cellular mobile telephone networks (where roaming may be thought of as a particular example of an IN service tailored to a specific situation).

In line with standard practice in telecommunications, a new protocol

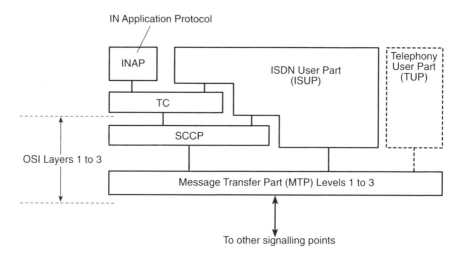

Figure 7.5   The C7 protocol stack for IN

has been developed specifically for the Intelligent Network. This is the so-called Intelligent Network Application Part (INAP), which may be considered part of the C7 protocol suite. Looking upwards, INAP interfaces directly with the Service Logic Execution Environment of the SCP. The Transaction Capabilities Application Part (TC) supports TC Users such as INAP (and MAP, the Mobile Application Part). It provides the OSI Session layer service together with dialogue control and is responsible for managing communications with remote TC Users.

INAP defines the C7 signalling messages relating to IN services and the functions and interactions they cause (in the form of finite state machines). INAP is in turn defined in terms of Abstract Syntax Notation 1 (ASN.1) making it independent of the computing platform and portable to any processing environment. This is an important consideration in the quest for IN products that will actually work with each other and in providing network operators with a means of enhancing their systems 'in-house' rather than being continually dependent on the manufacturers. INAP, like SCCP, is closely aligned with standards from the ISO—it is based on the OSI Remote Operations Service Element (ROSE).

One last point before moving on is to note that the Signalling Connection Control Part (SCCP) can ensure that any IN messages destined for a failed SCP are automatically re-routed to an operational one.

7.3   EXAMPLES OF IN SERVICES

Probably the best known example of an IN Service is Freephone (0800 in the UK), which we will use here to illustrate the main ideas of the Intelli-

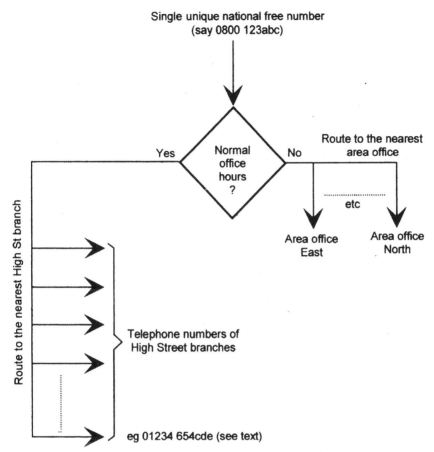

Figure 7.6 Typical Service Requirement of Large Insurance Company

gent Network and to introduce another IN network element, the Intelligent Peripheral (IP) mentioned above. The example given here is built around a hypothetical large insurance company which has branches in High Streets up and down the country, six Area offices each dealing with the administration of the High Street branches within their respective geographic areas, and a national Headquarters office in the capital.

A typical requirement of such a company, illustrated for clarity in Figure 7.6, would be to have a single, unique telephone number covering the whole country, such that:

- during normal office hours calls to that number would be routed to the High Street branch nearest to the caller;

- out of normal office hours calls are routed to the Area office covering the caller's location;

- when the Area offices are closed calls should be routed to the Headquarters office where they would be handled by the company's automated call handling system;

- the calls should be free (to the caller).

All of these requirements can be satisfied by using the established Free-fone service with the company being allocated an easily remembered number, say 0800 123abc. In reality, the branch and Area offices will have a variety of uncoordinated telephone numbers. The list of translations from 0800 123abc to the appropriate branch or Area office telephone number is stored in the central IN database, that is, the SCP, together with the time-of-day, day-of-week, and day-of-year routeing schedule.

So, how does this all work? Let us say that it is 09:31 on a normal Monday morning and a customer (or potential customer) of the company dials the company's national number, 0800 123abc. Though it is not necessary, we will assume in what follows that both caller and company are ISDN-based, so ISDN access signalling (ITU-T recommendation Q.931) is used at both ends of the call as shown in Figure 7.7. The caller's telephone number is 01234 567abc.

The basic call control process in the serving SSP, by reference to its Trigger Table, recognises the 0800 code as involving an IN service. Basic call control is then suspended and the SSP sends a C7 signalling message to its SCP giving the dialled number, 0800 123abc, and the caller's telephone number, 01234 567abc (the CLI). By reference to its routeing schedule for that particular 0800 number (123abc) the SCP knows that for that time-of-day, day-of-week, and day-of-year the call should be routed to the High Street branch nearest to the caller. By reference to the CLI the SCP knows that the telephone number of the nearest such branch office is

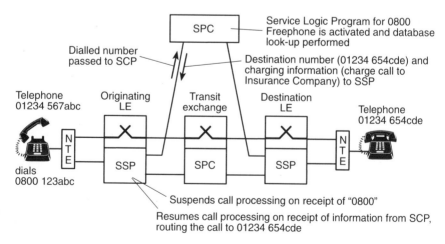

Figure 7.7 The IN Freefone Service

01234 654cde (the caller's nearest branch office may not have the same area code, it depends on the geography).

The SCP then returns a CCSS7 signalling message to the SSP advising that the actual destination number for the call is 01234 654cde and that the caller should not be charged for the call. On receipt of this message the SSP resumes basic call processing and the call is routed through the network to 01234 654cde in the usual way, and the call is charged to the insurance company.

Freephone with User Interaction

Let us suppose now that the insurance company takes over a financial services company and wants to incorporate the associated investment business into the existing company structure and processes. A new requirement is to direct customers' calls to the right sales team, either for insurance or for investment business. It is important to keep both types of business strictly separate because they come under different regulators. Given this constraint, we now need to ask the caller to indicate the nature of the inquiry. So, at the appropriate point in the call, we have the IN to send the caller an announcement along the lines of 'If you want investment services please enter 2. For insurance services please enter 3'.

To do this we need to modify the Service Logic Programme in the SCP to reflect the new requirement. And we need an additional network element, the Intelligent Peripheral (IP) as shown in Figure 7.8, to provide the customised announcements to the caller and detect any of the MF digits that are entered.

The Intelligent Peripheral provides 'specialised resources' such as customised announcements, concatenated announcements, MF digit

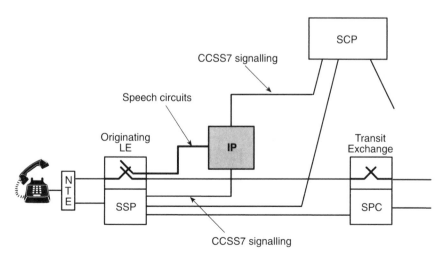

Figure 7.8   The Intelligent Peripheral (IP)

collection, signalling tones, and audioconference bridges. In due course no doubt it will also incorporate capabilities such as voice recognition to simplify the user's interface with the network. It is linked via standard traffic circuits to the switch so that it can be connected to the right user at the right time. It also has C7 signalling links to both the SSP and SCP.

There is a design choice of controlling the Intelligent Peripheral directly from the SCP or indirectly via the SSP. Clearly, with such comprehensive capabilities the Intelligent Peripheral is going to be an expensive piece of kit, and it may serve more than one SSP depending on cost/performance considerations.

Figure 7.9 shows how the Service Logic Programme would be modified to meet the requirement for customer interaction. Again, it is 09:31 on a normal Tuesday morning, and the caller dials 0800 123abc, the national number that we have set up for our insurance/financial services company. The call set-up proceeds as before up to the point at which the SSP suspends basic call processing and sends the C7 signalling message to the SCP containing the dialled number, 0800 123abc, and the caller's number, 01234 567abc. But this time, the Service Logic Programme in the SCP specifies that interaction with the caller is needed to complete the call. Specifically, it requires an announcement to be returned to the caller saying 'If you want investment services please enter 2. For insurance

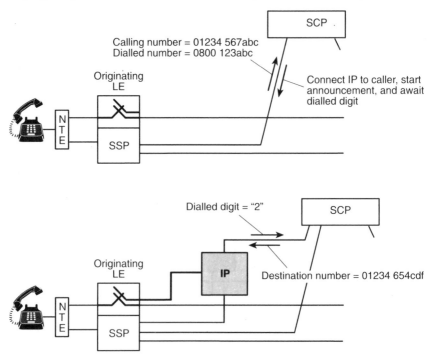

Figure 7.9   An IN Service with User Interaction

service please enter 3'. It requires a digit to be received by the Intelligent Peripheral before it will proceed.

The SCP therefore sends a C7 signalling message to the SSP asking it to connect the Intelligent Peripheral to the caller and start the announcement, as shown. On receipt of the digit entered by the caller, say '2', the Intelligent Peripheral sends this supplementary information to the SCP via the SSP which also releases the IP from the caller. The SCP responds with a signalling message advising that the destination number for this call is 01234 654cdf, the number of the nearest High Street branch's investment services team. At this point the SSP resumes basic call processing using 01234 654cdf as the destination number and the call is completed in the usual way.

Note that, in general, transit exchanges do not have SSP functionality (though they may). The power and flexibility of C7 allow functionality to be optimally located, so the same sort of cost/performance trade-off can be calculated as with exchange dimensioning.

Centrex, an example of a VPN

Centrex, which has been widely used in the USA for a number of years, is perhaps the best-known example of an Intelligent Network service. It provides a flexible alternative to the traditional private network and is, in effect, a Virtual Private Network (VPN). That is, it segments a part of a public network and reserves it for private use.

Traditionally a company's private voice network would consist of a number of Private Automatic Branch Exchanges (PABXs) interconnected by leased lines. Centrex does away with the PABXs. Instead, the functions provided by the PABXs are substituted by the same ones from the service provider's IN.

At each customer site access to the public network—the IN—is provided by means of one or more multiplexers which concentrate the site's telephones on to IN access circuits as shown in Figure 7.10. These multiplexers act as concentrators in that there are fewer voice circuits between the multiplexer and the IN than there are telephones on the site. It would be a very rare event for every telephone to be in use at the same time, and the size of the access circuit group is dimensioned using well-known teletraffic principles.

Centrex offers the corporate customer a similar service to the traditional private network of PABXs, including customised numbering schemes (typically of seven digits), but tends to be more flexible. For example, the numbering scheme can include small remote offices having only a few telephones, or perhaps only one, as shown in Figure 7.10. Telephones in such remote offices would have both a seven-digit number (say) and a standard PSTN or ISDN number. The jargon for this is that telephones can have both on-net and off-net numbers.

In addition Centrex offers uniformity of features in supplementary

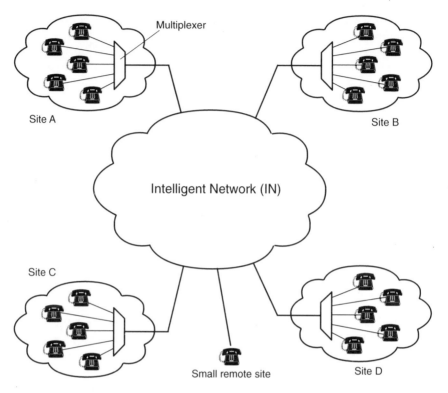

Figure 7.10 Centrex

services (such as call diversion, ring-back-when-free etc.) across all sites, including small remote offices. With networks of PABXs such uniformity is rare if different makes of PABXs are involved as is usually the case. Historically the most advanced features have been available in PABX implementations before they have been available in Centrex offerings but this is now changing as IN implementations mature and Centrex is often in the lead.

Green field situations where a company could choose between a private network of PABXs and Centrex are very rare, if they exist at all. Every company has a history and therefore a legacy, usually in the form of a PABX network. This cannot be simply discarded, at least not econ-omically, and moving operations in one step from a PABX network to Centrex would not in general be acceptable. It would put too many eggs in an untried basket. So in practice Centrex tends to be used in combination with PABXs in what is known as a 'hybrid' network as shown in Figure 7.11.

This requires additional functionality in the IN to integrate the PABX functionality with the Centrex service. Any Centrex service that does not support hybrid working will not sell very well. A possible drawback of a

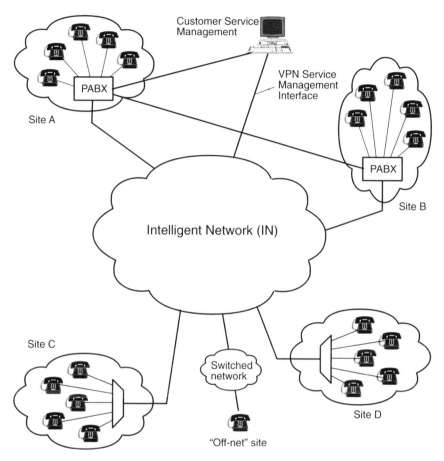

Figure 7.11 A Hybrid Network

hybrid network is that there may be some loss of transparency in the supplementary services as implemented by the PABXs and by the IN. Over time, as PABXs reach the end of their economic life, if the Centrex service proves to be reliable it is likely that a company will migrate towards a wholly Centrex solution. Experience in the USA is that most large companies use more than one Centrex supplier in order to avoid becoming locked into one supplier and to keep the Centrex service providers 'on their toes'.

There are basic economic differences between private networks and Centrex. Private networks have high capital costs for the infrastructure but usage costs are fixed and known in advance. With Centrex however the capital costs are small but running costs include a usage element that may be difficult to forecast. However, as we have seen, Centrex offers considerable agility to track changes in a company's operating and trading environment, whilst the high capital investment bound up in a traditional private network tends to militate against this. It should also be remembered that large companies tend to be multinationals, and the scope for cost savings on international leased lines can be considerable. The main VPN suppliers tend to have international coverage, often based on a consortium of service providers.

At this point we will change terminology from Centrex to VPN since the features we are going to discuss tend to go beyond those traditionally associated with Centrex. But the distinction is not important and in practice is likely to reflect marketing considerations as much as technical ones.

Access to VPN services may be obtained indirectly via one or more switched networks as shown in Figure 7.11. The terminology is that sites directly connected to the IN are 'on-net', whilst those with indirect access via another network are 'off-net'. Indirect access may be implemented in a number of ways. An off-net user may dial a pre-allocated access code, typically of four digits. This would route the caller to an access port on the IN, which would then implement a dialogue with the caller to obtain an authorisation code. After receipt of the authorisation code the caller would be treated as a regular part of the VPN (for that particular call: the authorisation code is used on a per call basis). Freephone access can be regarded as a particular case of a pre-allocated access code. Indirect access can be used to include public payphones and mobile telephones as part of a VPN.

Clearly telephones that gain indirect access using an authorisation code may also be used independently of the VPN (simply by omitting the access code). Another option, however, is to use 'hot-line' access whereby a remote telephone is dedicated wholly to the VPN. When the handset is lifted on such a telephone it is immediately connected through the access network to an access port of the VPN. In this case an authorisation code is not needed since the Calling Line Identity of the caller is passed to the VPN by the access network (in effect there is 'point of entry policing') and the telephone is usually regarded as on-net.

One last point that we illustrate in Figure 7.11 is that of a customer management of their VPN. In the traditional private network of PABXs, the company had complete control of their network. This may be seen as an undesired burden or as an important business control. But in any case it creates an expectation on the part of the company that it should have a degree of direct control over their network, and with the increasing competition between VPN service providers, the degree and ease of control offered to the VPN customer is an important differentiator. Typically a VPN customer will be offered remote control of a number of service aspects including:

- the numbering plan;
- authorisation codes and passwords;
- call routeing (for example, where a call is routed partly in the VPN and

partly in another network it may be important to control the VPN routeing to minimise call costs in the other network);

- call screening (that is, barring of international calls or premium rate services, perhaps with over-ride authorisation codes).

In addition the VPN customer would typically be provided with on-line access to reports including:-

- network costs—the customer wants no surprises in budgeting;
- network performance—to check that service level agreements are being met by the service provider;
- details of usage, such as calls made out of normal business hours;
- network traffic reports, typically produced daily, weekly, monthly, on-demand, or whenever preset thresholds have been exceeded.

General
Summarising all this, the basic ideas of the Intelligent Network are:

- to separate basic call control from customised aspects of call control;
- to do basic call processing in the Service Switching Point (SSP);
- to do the customised aspects of call control centrally in the Service Control Point (SCP);
- the detailed information relating to a corporate customer's service is held in the SCP (usually);
- the Service Switching Point is a modification of the existing exchanges;
- the Intelligent Peripheral (IP) provides specialised resources.

The whole point of this centralisation of intelligence is to ease the otherwise intractable problem of creating and managing the customised services for the corporate customer. In this way we can realise the full potential of the stored programme control switches around which the telecommunications networks is built to provide services tailored to the requirements of individual customers quickly, flexibly and reliably. To achieve this in practice additionally requires effective means for creating and managing these individually tailored services.

7.4   SERVICE MANAGEMENT AND CREATION IN THE IN

We have already seen in Chapter 3 that the management of networks is a key part of ensuring their operational effectiveness. The specific concern with an installed IN is to manage the services that are provided. So, we need to attend to those issues associated with the Service Management Layer of the Telecommunications Management Network (TMN) model introduced earlier. Specifically, service management in the IN entails:

- Adding new customers and services;
- Synchronising changes to service data and Service Logic Programmes across the SCPs (as we have seen, there will always be more than one, possibly quite a few);
- Administering the database(s);
- Reloading service data and software following an SCP failure and managing its return to service;
- Bringing new SCPs into service;
- Network and service surveillance.

This list is certainly not exhaustive, but does cover the main areas.

Good service management requires a good Service Management System (SMS). This is crucial for the service provider to remain competitive as the Service Management System should be the subject of timely upgrading and enhancement to reflect new demands, new threats and new opportunities.

The place of the Service Management System in the scheme of things is shown in Figure 7.12.

Historically companies will generally have invested heavily in their own private network infrastructure. Their control and management of this network infrastructure has traditionally been both comprehensive and direct. If they are to be persuaded to forsake their private network for a VPN approach they need to feel that they are still in control. An important element of the SMS therefore is the management interface and capability it gives to the customers. They want, and are used to having:

- Up-to-date information on performance;
- The ability to change telephone numbers within 'their' network;
- The ability to change access authorities, such as barring international calls or calls to premium rate services.

But customers' on-line access to the Service Management System has to be very carefully controlled to ensure that they cannot, deliberately or un-wittingly, affect somebody else's VPN (or indeed, the service provided to the 'public'). Firewalls are therefore used to prevent customers from getting access to any capabilities they are not entitled to (or have not subscribed to).

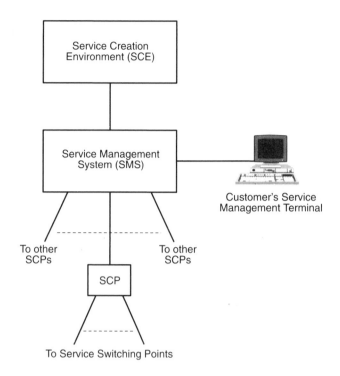

Figure 7.12 Service Creation and Service Management in the IN

Service Creation
We have already suggested that the speed at which a service provider can produce a customised solution to corporate customers' needs is an important aspect of his competitiveness. An effective Service Creation Environment (SCE)—basically the set of tools used to create and test new or customised services—is therefore a key requirement for success. And once you have made the 'sale' you cannot afford to lose it because of poor operational performance. There is also considerable scope for a flawed IN service to wreak havoc in the public services provided on the same network platform. So service creation needs to be not only fast, but also robust and accurate.

The Service Creation Environment is likely to use object-orientated methods, a powerful graphical user interface, and an Application Programming Interface (API) that reflects the Service Logic Execution Environment. Ideally, it will support the complete service lifetime, embracing requirements capture, service specification, service demonstration, design and development, service trials, software release control and deployment, and in due course service termination. Figure 7.13 shows a typical service creation process.

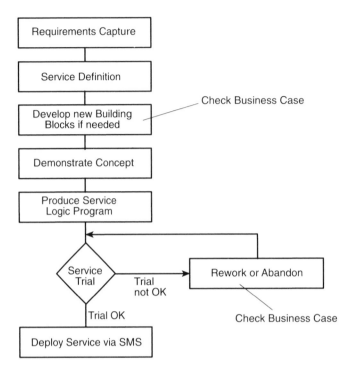

Figure 7.13 The Service Creation Process

7.5 CENTRALISED OR DISTRIBUTED INTELLIGENCE?

Our treatment of the IN has been based on centralised intelligence, since in practice it is virtually impossible to control a distributed intelligent database spread over hundreds, or even thousands, of locations, at least when the service information is changing frequently. But the leading light in telecommunications standards, the ITU-T, have, in their wisdom, also created IN standards that embrace distributed intelligence. One of the reasons for this is that, despite the problem of managing such distributed intelligence, there are some situations where centralised intelligence just does not make sense. An obvious example is the Internet, where intelligence is vested in peripheral devices and the network itself is, in essence, a connectionless data service—each switch in the net takes local action based on the information in the packets it sees (Figure 7.14).

In the Internet, there is no real sense of a call being set up. Once logged on, a user can happily send half a dozen mail messages to different addresses and surf lots of unrelated web sites. So if we add centralised intelligence, we would have to make reference to it on a packet-by-packet basis—and this would simply take too long. The switching delay incurred

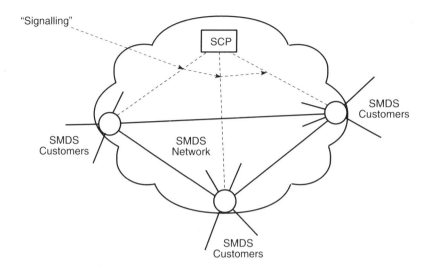

Figure 7.14 Internet intelligence

by an IP packet is typically be less than a millisecond. But it would take several tens of milliseconds for a routeing enquiry to be referred up to the central database and for a response to come back. This is clearly untenable.

Even so, some attention within the Internet Engineering Task Force (the source of the IP protocol) has been focused on adding 'central control' features to the Internet. For instance, a mechanism for 'tagging' IP packets (called MPLS, more of which in Chapter 10) has been devised and this would enable VPNs to be defined within the Internet, at the network level.

How might this work? The creation of an operational VPN for a small company requires entries in routeing tables of all the switches that serve that company. In practice this would be done by maintaining a central 'map' of the VPN and down-loading incremental changes to routers to reflect changes in service requirements. And this would reap many of the benefits of centralised intelligence whilst keeping the flexibility (and fast response time) inherent in the established model where intelligence is in the peripheral devices.

Returning to mainstream IN thinking, the ITU-T distinguish between a function and its location. Figure 7.15 summarises the main functional elements used to build Intelligent Networks and where they may be located.

Though we have not considered it in the above description, there is also a Call Control Agent Function (CCAF) which defines a subset of the Call Control Function (CCF) that may be implemented remotely from the CCF. Typically a CCAF would be located at a Local Exchange with the full CCF implemented at the nearest Transit Exchange. There is also a Service

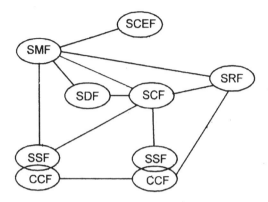

CCF   Call Control Function
SCEF Service Creation Environment Function
SCF   Service Control Function
SDF   Service Data Function
SMF   Service Management Function
SRF   Specialised Resource Function (The IP)
SSF   Service Switching Function

	SSF/CCF	SCF	SDF	SRF
SSP	Mandatory	Optional	Optional	Optional
SCP		Mandatory	Optional	
SPD			Mandatory	
IP	Optional			Mandatory

Figure 7.15 IN Functional Elements and their Location

Management Agent Function (SMAF) which defines parts of the Service Management Function (SMF) that may be implemented remotely from the SMF.

The mandatory entries in Figure 7.15 reflect the centralised view of intelligence as we have developed it for the IN. The optional entries embrace distributed intelligence.

7.6   SUMMARY

This chapter has developed the idea of separating the telecommunications network into a switched information subnet and a signalling subnet and has shown how this allows intelligence to be added to the signalling subnet. This fulfils the original promise of stored programme control and in effect puts the operator back into the network: after all, the most intelligent network technology ever deployed was the human operator sitting at a manual switchboard. In the very early days of telephones it was possible for the small group of operators looking after a small town to know everyone in the town who had a telephone and their business. If someone called the doctor, the operator would typically know that he was delivering Mrs Smith's baby at No. 16 London Road, and could redirect the call accordingly.

The separation of the switched information subnet and the signalling subnet was a feature of the manually-switched telephone networks of the nineteenth century. Indeed, the separation was not only logical but also physical. The switched information subnet consisted of the manual boards (panels of arrays of jack sockets into which the operator inserted the appropriate plug-terminated cords to effect the desired connection), whilst the signalling subnet consisted of the operators (who sat at adjacent exchanges and communicated with each other by means of 'order wires', the equivalent of the C7 Message Transfer Part). The customer's signalling interface with the network was the most natural imaginable—free-format speech.

It is interesting to note that the intelligence and flexibility of this control mechanism could be easily subverted and that this was what led to the development of automatic telephony. The story, perhaps apocryphal, is that the aforementioned Almon B. Strowger, an undertaker by trade and the inventor in 1898 of the first viable automatic switches for telephony, was motivated to design his switch because he found out that a rival undertaker had bribed an operator to divert calls, and thereby business, from Strowger to himself! Perhaps what this illustrates is that, whilst we may have the odd malfunction in the speech or data path, it is the signalling path in which the real power lies to create havoc, and to develop the power of this to fulfil the full potential of stored programme control requires a well-defined architecture if the havoc is to be avoided.

This chapter has:

- developed the notion of separating the signalling and switched information subnets;
- explained how intelligence is added to an established circuit switched network to form an Intelligent Network or IN;
- introduced the main ideas and terminology of the Intelligent Network;
- shown how the IN enables customised network services to be constructed.

The approach described in this chapter for implementing flexible network services is not the only one. It is the approach of the telecommunications community and has been designed to provide the flexibility and scalability needed by many large, corporate customers. Our last section in this chapter questioned whether centralised intelligence could be applied with the Internet, where intelligence has traditionally been vested at the edges of the network.

REFERENCES

Jabbari, Bijan and Gervais, Pierre (1994) *Intelligent Networks*. John Wiley & Sons.
Thorner, Jan (1994) *Intelligent Networks*. Artech House.
Venieris, Iakovos and Hussman, Heinrich (1999) *Intelligent Broadband Networks*. John Wiley & Sons.

See also ITU-T standards on Intelligent Networks (these are in the Q series recommendations). The reference for the ITU is given in Chapter 9.

8

The Internet and Intranets

If you're not on the net,
you're not in the know

Fortune magazine

If the Intelligent Network exemplifies central control in a connection-orientated environment, the Internet appears at the other end of the spectrum. In this chapter we are going to look at a network that is connectionless, where virtually no central control is exerted and where intelligence is vested in peripheral devices. The spirit that underlies the Internet is probably best summarised in the credo of its technical lead, the Internet Engineering Task Force, or IETF: 'We reject Kings, Presidents and voting. We believe in rough consensus and running code'.

Large and complex things like the Internet are always difficult to classify. The problem is exacerbated by the fact that it is distributed. The ancient story of a group of blind men encountering an elephant exemplifies the problem. The first man, on feeling its trunk, thought it was a snake. The second felt its tusk and declared it a spear. A third thought it had to be a tree when he felt its leg. It was only when they compared notes that they pictured an elephant.

So, rather than try to build a single picture of a network in operation, as we did for the Intelligent Network, we will describe the Internet from several different points of view in this chapter. Once we have a clear picture of how the Internet works, we move on to take a brief look at how the same technology is used to create communities of interest (extranets) and private virtual networks (intranets).

8.1 VIEWS OF THE NET

If the Internet looks different from different viewpoints, a first question has to be 'what is your viewpoint?' Are you a residential user with a computer and a telephone. Perhaps a business user looking for a corporate connection? Or maybe an Internet Service Provider seeking to serve a set of customers? You might even be a not-for-profit organisation tasked with the management of an Internet connection exchange? Whoever you are, your perception of the Internet will differ. Of course, we are privileged, because after we have seen a few key perspectives we will have a good overall impression of how the parts fit together. Those that we will use here are:

- The dial-in end-user
- The direct connection user
- The Internet Service Provider (ISP)
- The global Internet

To a first approximation, the first on this list is a domestic user, the second a business user, the third a commercial supplier and the last the underlying technology. In each case, we will explain the network options, hardware requirements and typical software on offer.

The dial-in end-user

It is only over the last few years that this type of user has grown significantly in numbers. The Internet, once the province of the University research fellow or IT professional, has been rapidly adopted for domestic and social use. In turn, this has driven the Internet service provision market to meet this demand.

The popularity of the Internet has grown from the availability of a single software application—the Web browser—and its logical network of information—the World Wide Web. The Web browser, the Internet addressing scheme, the UNIX file system, Java and the HyperText Mark-up Language (HTML) have been combined to build a powerful information addressing and viewing system. In essence, the World Wide Web has brought point-and-click usability to network computing with timely delivery of information, thus opening up access and availability to information on a global scale.

The first thing that we need to consider for the dial-in user is what we mean by dial-in. This means looking at the connection between the user and the rest of the world. In the vast majority of cases, this comes down to one of two options, both introduced earlier on: the Public Switched Telephone Network (PSTN) or the Integrated Services Digital Network (ISDN). The former is the familiar phone-socket-on-the-wall that you connect either your telephone or your modem to. The latter is a less familiar socket-on-the-wall that also carries voice traffic but, by virtue of being all digital, allows data to be carried without the need for a modem.

As far as a user is concerned, the main distinction between PSTN and ISDN is speed. The fastest modems operate at 56kbps whereas the ISDN is built around basic channels of 64kbps (and can deliver higher rate connections—up to 115kbps—with compression).

There are also differences in the peripheral hardware that is used to connect to the telecommunications network. In the case of PSTN, a simple modem is all that is needed. With ISDN there is a choice between a terminal adapter (a means of connecting your computer's communication bus to a communications service) or router (a means of connecting a network port to a communications service).

The various links in the ISDN access chain are shown in Figure 8.1 for each option.

If either a modem or terminal is used, there are two ways of connecting it to the PC. The first is via the serial port (the familiar 25 pin RS-232 interface on the back of the machine). The second is with an internal card which can either slot onto the bus inside or, in the case of the PCMCIA card (which is more relevant to laptops and portables), can be inserted in a small slot on the side, which then connects to the bus. Bus connected devices tend to operate much better since their performance is not hindered by external device interfaces, such as serial ports.

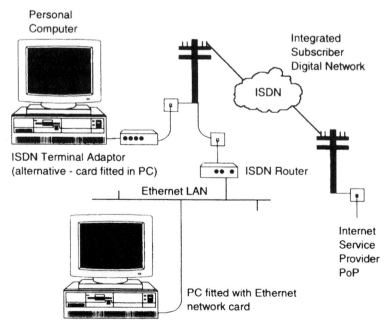

Figure 8.1 Dial Access options

Routers are a different story altogether. These devices can connect many PCs (e.g. together on a Local Area Network) to the Internet. If you have a router and you intend to connect to the ISDN with your PC, you will need a network card in the PC to create the network between the PC and the router. This set-up is more appropriate for the small office environment where there may be shared devices, such as printers, and other PCs on the network. With multiple PC connections to the Internet using LANs and routers there are many more things to consider, such as addressing and security, and these are covered in the following section.

Access to the Internet over the PSTN is achieved by the modem that transmits signals from the software on the PC to another modem at the ISP's Point of Presence (PoP). These digital signals are converted by the modems for transmission over analogue telephone circuits. When the signals break out into the ISP's PoP the signals are converted back into digital data and sent on their way across the ISP network to the global Internet.

The software required for gaining access to all the services that the Internet has to offer can be quite involved. It can appear as though many programs with little apparent function have to be present. The confusion eases once the basic building blocks are explained. In terms of task performed, there are really just three program types:

1. The applications or programs that work over the Internet (e.g. Web browser, FTP etc.). These allow you to carry out some useful tasks (retrieve files, access information etc.). They provide the familiar face of the Internet.

2. The communications software for communicating across the Internet. This allows you to talk to other machines and resources on the net. An overview of the underlying protocols that make the net work is given at the end of this chapter.

3. The communications software for transmitting the Internet communications protocol across the telephone (or ISDN) line. This enables you to connect over a variety of bearers. For the dial-in user there are two main mechanisms—point to point protocol (PPP) and serial line interface protocol (SLIP). Both operate in background to allow the Internet Protocol to work over the phone link as it would through a network connection. PPP is the newer, and faster, of the two.

Each building block can be envisaged as sitting in a stack and cooperating to effect a transparent connection to the network. In this context, the word 'transparent' is used to indicate that there is software to shield the user from the vagaries of the access path they are using; they can see to remote service or application they need with no (network specific) impediment.

The stack picture is not actually a real picture of how the software is

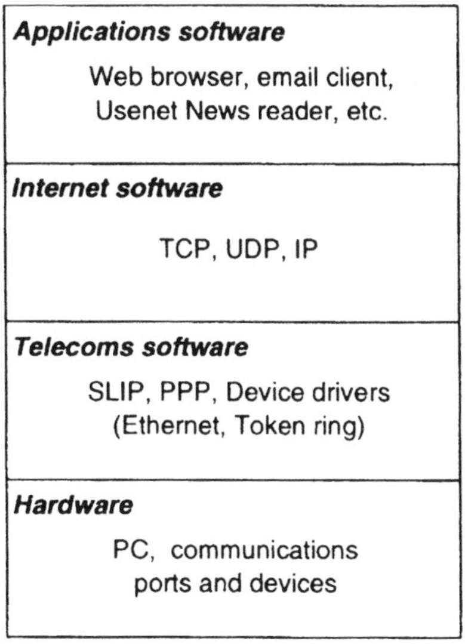

Applications software Web browser, email client, Usenet News reader, etc.
Internet software TCP, UDP, IP
Telecoms software SLIP, PPP, Device drivers (Ethernet, Token ring)
Hardware PC, communications ports and devices

Figure 8.2 The PC Software stack

organised in the PC, but it does represent a convenient structure that illustrates how one piece of software interfaces to another. It also demonstrates how important it is for all pieces of software to be present before it all works, to get one application talking to another, both ends of the stack must have all the foundations in place.

In reality, there are other software components, used for driving the modem or ISDN terminal adapter. In addition, the Internet Service Provider (more on them shortly) may provide components of software for registering your subscription to their service, assigning bills and so on. The important point, though, is that some software you use directly; some takes care of the complex and diverse structure of the network you are using. The former may be more enticing, but both need to be in place.

The direct connection user

In contrast with the visiting dial-up user, the direct connection user is always present. Typically, this is an academic or business user, often someone who uses the Internet routinely as a part of their job. A direct connection between the customer and the ISP is available 24 hours a day, 7 days a week (or so we hope). Rather than paying for the connection on a per minute or per hour rate, the connection becomes a permanent feature that is rented by the customer from a telecommunications supplier (often a third party to the ISP). This agreement attracts a fixed rate tariff that is often dependent on the distance from the customer to the ISP's nearest point of presence (PoP).

Table 8.1

Method	Typical Bandwidth	Explanation/comment
SMDS	Up to 34Mbps	Packet-switched connectionless network, available primarily in the UK
Frame Relay	Up to 2Mbps	Packet network, globally available
X.25	Up to 9.6Kbps	The original packet-switched standard, old and slow but cheap, ubiquitous and reliable
Private Leased Circuit	Multiples of 64Kbps and 2Mbps	A dedicated point-to-point circuit. Termed Kilostream or Megastream in the UK.

A direct connection can be implemented in a number of ways with a number of different technologies. Some have been around a bit longer than others and so the functionality or bandwidth may be limiting factors.

Table 8.1 gives a short summary of the available bearer services in the UK.

It should be clear that there are all strengths of bearer services available, from the occasional back-up line through to the high capacity data high-way. Yet faster and better data networking technology is on the way in the form of Asynchronous Transfer Mode (ATM) and ADSL (both explained in earlier chapters) we'll take the emergence of the superhighways as given.

In each case, the suitable capacity circuit is used to connect a network at the customer's premises to the ISP's PoP via a router. To all intents and purposes, this circuit belongs to the customer. Again, we can draw a picture of the various links in the access chain—the two alternatives (direct LAN connection and remote LAN connection, respectively) are shown in Figure 8.3.

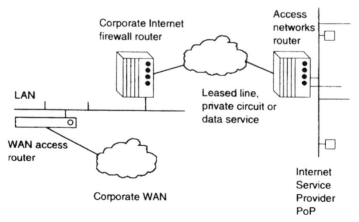

Figure 8.3   Private circuit access to the Internet—the LAN and WAN options

As previously mentioned, a router is a device that connects networks together in order to route data traffic between them. Typically, Local Area Networks connect many PCs, workstations and other devices together. As well as allowing local resources to be shared, they can provide access to the Internet for a large community. PCs that are connected in this way will require a network card (and software) to be installed. The software configured on the PC (TCP/IP) will need to be configured with an address (an Internet address) and the address of the router (the default gateway address) that connects the LAN to the Internet.

So much for the physical connections. Because the directly connected user is often part of a larger organisation, there are additional subtleties we now need to know about. One of the more notable is that there has to be some way of telling one user of a direct connection from another who uses that same access route. The dial-up user's name and phone can usually be regarded as one and the same thing. When a number of people are connected to the same LAN you need to be able to tell them apart in just the same way. So addressing is important.

If the network they share was already in place prior to connection to the Internet it may have a block of IP (Internet Protocol) addresses already assigned to it. These are addresses that are usually supplied by the ISP and registered with the Internet addressing authorities (the InterNIC or RIPE). Each device on the network requires an IP address before it can be connected and addressed on the network. Without an IP address, a device, such as a host, is not addressable and hence useless.

When IP was created the Internet Network Information Center (Inter-NIC) was set up to oversee the allocation and registration of these addresses. However, because the specification for IP was well documented and public, a considerable community of system administrators across the globe took it upon themselves to use blocks of addresses as they saw fit. The assumption at the time was that the corporations that employed these networks would never want them connected to the Internet for anything other than email, therefore multiple conflicting addresses would never become a problem.

Because email has its own protocol and architecture, it does not impose full Internet access on all participants on the network. It is an application that routes its traffic between mail hosts using SMTP (Simple Mail Transfer Protocol) and does not actually rely on the presence of IP at all. Other protocols such as UUCP (Unix-Unix Copy Program) can equally well be used to transfer SMTP between hosts.

There are two ways of accessing a mailbox on a host, either through terminal access (which does not have to use IP) or using a POP (Post Office Protocol) mail client. The mail host therefore, can act as a firewall between the user community and the Internet, since Internet email only needs to be delivered to the doorstep of a private network before being routed internally (safe from any address conflicts) on a private network.

So, thought many planners and administrators, there was no need to register IP addresses: the only use of the Internet would be for email transport and this did not require full access. We now know that this is a false assumption with greater exploitation of the Internet being sought. The legacy is of unregistered IP addresses, the potential problem is address contention and the action a rethink of the addressing policy used.

And Internet addressing is not only a problem for organisations that have an unregistered scheme and want to pursue a full Internet connection; the address space itself is running out. This is a physical limitation on the number of unique addresses that the Internet can hold. The addressing structure (explained later in this chapter) is split into a number of classes that sub-divide the address space. Its limitation is set by its size—32 bits—with each sub-division capable of addressing between a few tens and many millions of devices.

Estimates of when this address space will be fully allocated vary according to the text of the day, but it will not be long—a few years at the most. The non-deterministic nature of the problem is illustrated (and compounded) by the structuring of addresses into classes. The higher classes can be used to address many more devices than the lower classes. Thus when an organisation that registered a class A address relinquishes it to InterNIC in exchange for a class B or C (because the several million devices that a class A can refer to are more than enough for any one) the life of the address space is extended. Trouble is that no-one really knows how much spare is left in the system or how long it will last.

Needless to say, such a fundamental issue is not being ignored and an upgrade to IP (known as IPv6 or IP next generation) is well under way. Again, this is briefly explained towards the end of this chapter.

Coming back to the mainstream of this section, it is tempting to reflect on the motivation for a direct connection. After all, the cost of bearer capacity and complexity of addressing must have some recompense. So why would you choose a direct connection to the Internet?

Bandwidth is one major consideration, and access for a large community of users another. Some organisations (and individuals) perceive value in having a presence on the Internet, as opposed to just gaining access to it. The Internet, unlike the corner shop, is open all hours and so the value of the Internet presence is in full global availability 24 hours a day. This method of operation is not possible with dial-in connection using the PSTN or ISDN (at least, not economically). What we mean by presence is a World Wide Web site where information served from the site can act as a shop-front, front-door, advertisement or channel-to-market. Electronic Business, in effect.

The ISP will provide a number of ancillary services as part of a direct connection to the Internet. Whether these are bundled in the cost of connection, sold separately or offered free of charge will depend on the ISP in question. These services include, but are not restricted to:

- Domain name registration
- Internet address allocation
- Usenet news feed
- SMTP mail relay
- Managed router

Taking each of these in order

- Domain name registration involves choosing a name that defines the sub-network that you create when your connection to the Internet is in place. The ISP will register your chosen name with InterNIC or RIPE to identify the Internet addresses that you either have in place already, or are assigned to you by your ISP. The domain name is not essential but it is the component of the Internet that makes it more readily usable—a sort of on-line brand name. This is explained later in the chapter.
- Internet address allocation will only be relevant to you if you do not already have a registered address space. In this instance, the ISP will generally register an address space for you
- Usenet news is the Internet bulletin board. Originally developed by UNIX developers to share email on the development of the UNIX operating system, it has grown into an all-purpose bulletin board that carries thousands of discussion groups. The bulletin board architecture is that of a federated hierarchy where articles posted to a group on a host are replicated across the Internet to all other Usenet hosts that hold that group. There is not a single host that holds the bulletin board. Typically, ISPs maintain a Usenet host for their dial-in customer base and expect direct connection customers (especially where large corporate communities are involved) to maintain their own. Hence the news feed that extends the communicating news hosts model into the private sub-domain.
- An SMTP mail relay is a standard feature for an ISP and is used to route email traffic from the ISP customer base to the Internet. Unlike national email services based on the ITU-T X.400 standard there is generally no cost associated with the forwarding of SMTP mail in this way.
- ISPs may also offer a managed router as part of the direct connection package. This can be important for those that do not want the management overhead of another network box to maintain and the added complexity of creating a firewall between a private network and the Internet.

Firewalls are an essential security feature of any network-based connection to the Internet. The firewall primarily offers protection from hackers but can be used to restrict user access to the Internet on a time-of-day or site-of-special-interest basis. These tools are becoming increasingly important in the enforcement of corporate policy on such matters.

It should be clear that the direct connection user has an altogether different view of the Internet from the dial-up user. Their level of investment is higher as is, most likely, their level of dependence on and service from the Internet. Also, the direct connection customer is usually obliged to know a considerable amount. They will not generally be provided with Internet software and will have to make provision for the care of their network resources. Part of this will be the management of routers and, in more sophisticated cases, the provision of firewall configurations that involve multiple routers, workstations, servers and networks.

In short, the direct connection is a more responsible user option, with more on offer.

And so to the other side of the coin, the supplier

The Internet Service Provider (ISP)

The ISP is the provider of Internet connection services. They retail access to end-users, both dial-up and direct. The rise in popularity of the Internet has been reflected in the growing number of ISPs: from around 10 or so in the early 1990s to over 200 by the late 1990s in the UK alone. Needless to say ISPs vary considerably in size and capability. Some are major enterprises, some are small cottage industries operated by one or two enthusiasts. It can be difficult to choose between them. To do so, you should have a clear idea of what you really need and what the most important differentiators are.

There are seven key criteria against which ISPs can be assessed, the most fundamental yardsticks by which the ISP can be measured

1. Network capacity and topology
2. Number of Points of Presence (PoPs)
3. Range of access (connection) classes
4. Global interconnection
5. Dial-in services
6. Customer support services
7. Web space leasing (rent-a-web)

For most people, cost and quality of service will play an important role (although many offer free service). So any calculation to determine overall value for money would balance these against the intended use (i.e. importance of or reliance on) the Internet connection.

Different technical and operational requirements mean different importance attaches to each of the criteria.

1. Network capacity and topology

Network capacity and topology will affect the performance and quality of service that is perceived by the customer of an ISP. Capacity (or bandwidth) sets the theoretical limit on the number simultaneous connections to the network. The topology, and associated routeing policies, will determine how data is routed across the network according to traffic volume and availability of network paths.

Because the Internet is made up of many interconnected networks, each with their own capacity, topology and routeing policies, not to mention loading on the network from an indeterminate customer base, the same sort of traffic calculations that can be carried out for a telephony network are not possible (Figure 8.4).

In practice, ISPs work out the number of customers they can support on a PoP and include an arbitrary utilisation factor to support more customers than bandwidth available to those customers. In principle, this is the same as using the Erlang B formula to calculate the right amount of equipment for a given grade of service.

2. Number of PoPs

The number of PoPs an ISP has may influence a decision to choose one ISP over another. For example, an operation that wants to open an 'Internet

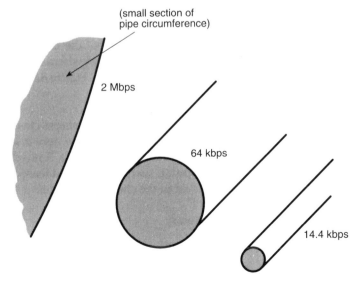

Figure 8.4   Relative bandwidths

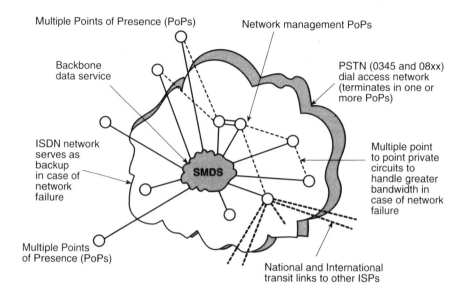

Multiple Points of Presence (PoPs)

Network management PoPs

Backbone
data service

PSTN (0345 and 08xx)
dial access network
(terminates in one or
more PoPs)

ISDN network
serves as
backup
in case of
network
failure

SMDS

Multiple point
to point private
circuits to
handle greater
bandwidth in
case of network
failure

Multiple Points
of Presence (PoPs)

National and International
transit links to other ISPs

Figure 8.5   Multiple PoP ISP network

Cafe' in every major city may choose the ISP with PoPs in every city, since the cost of a few long leased lines are typically more expensive than many leased lines of short length. Alternatively, a roving salesman with Internet dial access may want access from a list of cities that coincide with his territory; the evaluation here is that UK national calls cost more than local calls for dial-in access. An alternative in this scenario is that the ISP may provide a single local-call number that can be dialled from anywhere in the country. In this case the number of PoPs is not so important (Figure 8.5).

3. Range of access (connection) classes
An ISP may have many PoPs but does each one have the right access class for you? There are many access classes that define the basic network access needed to gain access to the ISP's PoP. Most ISPs offer a limited selection, usually those that prove most popular, such as PSTN and leased line. A comprehensive list (touched on earlier in the chapter) includes:

- PSTN   Public Switched Telephone Network

- ISDN   Integrated Subscriber Digital Network

- PLC   Private Leased Circuit (or Leased Line)

- FR   Frame Relay

- PSDN   Public Switched Data Network (typically based on the X.25 standard)

- SMDS   Switched Multi-megabit Data Services

4. Global interconnection

If timely access to globally distributed information is a concern, you will want to know how many connections the ISP has to other ISPs and the bandwidth of each. Unfortunately this information is something that ISPs do not often advertise but they will generally explain their network topology, connectivity and capacity when pushed.

Typically, if an ISP has multiple interconnections they will almost certainly implement higher bandwidth connections to handle the volume of traffic they would expect across those routes. Additionally if an ISP has multiple connections, if one or more were to fail then there would be alternative paths over which to route traffic, thus making the network more resilient with higher availability (Figure 8.6).

5. Dial-in services

As we have already indicated, Internet Dial Access presents the ISPs with a major growth area. There are several reasons for this. First, the unit cost of home computing has fallen to average income affordability and the home PC has become accepted in the domestic market as a multi-media device, capable of integration between telephone, television, hi-fi stereo

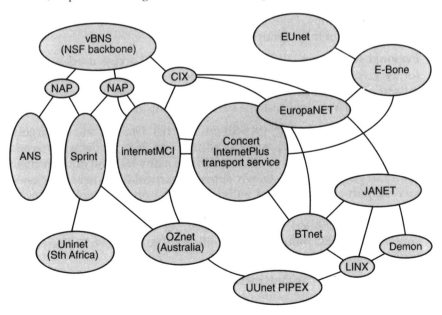

Figure 8.6   The global interconnection of ISPs

system, video and even digital camera. Second, the cost of dial access has fallen dramatically due to lower phone charges and a raft of free Internet access offers. With cost out of the equation, ISPs offering dial access can be discriminated by:

- **Software bundle** Does the service provide free software (web browser, dialler, TCP/IP stack, on-line registration, mail client, usenet news client, FTP client, Ping, Chat client and so on) or do you have to pay for it? There are obvious advantages in receiving a bundle for free (ready licensed your use, not public domain). It means that you will know that the software will work and that if you have any problems, the customer service desk will be able to help you out in the shortest possible time (albeit using a premium rate call). There are advantages to the ISP giving away software. Software can be 'hard-wired' to only work with the ISP network and so software licensing can be accurately accounted for.

- **Email** Does access to the Internet come with an email address (if so, how many and how memorable). An important consideration here is the protocol used to access the mailbox. There are two protocols in common use, SMTP and POP (these are explained later) and each one has some advantages over the other. POP or POP3 is most appropriate for dial-access.

- **Usenet news access** The Internet has a 'bulletin board' that is structured into discussion groups, known as Usenet News. There are over 10,000 such groups. Does the ISP provide free access to Usenet and, if so, how many of these groups does it carry?

- **Personal web pages** It is popular to offer dial-access users a facility for the creation of their own World Wide Web 'home page'. Some offer a fixed storage, such as 5 Mb, others are unlimited. If this feature is important for you, look into the various benefits offered.

- **Customer support** The growth in Internet Dial Access is largely fuelled by non-technical and novice users of PCs, modems, network software and applications software. Where many ISPs fall down is their ability to offer quality customer support, particularly where the customer support can be reached without hanging on the telephone for an hour or two. Customer service of this nature is not only expensive to provide but a limited resource in itself. At least one major software supplier to the PC industry has significantly scaled down its customer service operations for these reasons.

- **Charging models** An Internet access fee can then either be time based, a fixed fee or a combination of the two. Many ISPs will not make any usage-based charges because they want to reduce what is known as churn, or the cessation of service from a customer. Since customer registration and cessation both cost the ISP in time and resource, it is the aim of the ISP to retain its customer base and limit churn. Some are free, some will levy a subscription (usually for additional services)

In countries where telecommunications are not so stringently regulated, telecoms service providers are able to bundle tariffs for Internet access with the telephony bill. The cost of Internet access is significant in such instances as the additional call revenue alone can support the ISP business.

- **Number of PoPs and contention ratio on each PoP** ISPs provide many modems within a PoP, but not one for each customer and the basis of sharing modems between customers dialling into a PoP is first come, first served. ISPs will work out the number of customers they can support on a PoP in this way and work to a *contention ratio* that will enable them to support more customers than modems. This calculation increases the cost-effectiveness of the PoP and if managed correctly will not significantly affect service availability for the customer. There is not the same level of precision here as with telephone switches but anything higher than 15 or 20 to 1 should be avoided.

6. Customer support services

Customer support should be a differentiator with which to assess ISPs without getting bogged down in jargon and technicalities. If an ISP does not answer the phone, or can answer your questions then they probably cannot offer you a suitable service. This will apply to all customer types, corporate, small business or domestic.

7. Web space leasing (rent-a-web)

Some ISPs recognise that the need to connect to the Internet has arisen from the need to be present on the World Wide Web. Some may also contract with design agencies and software houses to design and build your Web presence for you. A complete package is often preferred and obviates the need to build specialist Website design and management skills in-house. There are benefits to both the customer and the ISP in this situation, especially where the Web site has a high profile, a significant customer base and requires expensive bandwidth to support concurrent access. This is bandwidth that the customer does not have to pay for directly, instead, they pay for a proportion of the bandwidth supplied to the server infrastructure. By balancing a range of Web sites across a server infrastructure in this way, the ISP can optimise its use whilst attaining an economy of scale largely unavailable to its customer base.

One adjunct to renting Web space is on-line advertising. Some organisations are now making considerable capital from their Web presence: the number of people accessing some sites makes it worth paying to share the presence. Time-Warner is one of a number of companies successfully exploiting a new advertising medium.

When cost, reliability, ease of use and preference are factored in with all of the above criteria, it should be clear that you cannot rate ISPs on a simple linear scale. The variety of uses that people put the Internet to and amount of reliance that they vest in it mean that the ideal ISP for one will be a poor choice for another. It is very much a matter of matching what is needed to what is on offer.

The global Internet

The Internet is a global phenomenon that has reached almost every country. In each region there are one or more ISPs that offer connectivity to their customers and also to their competitors and allies. The Internet would not work if ISPs did not interconnect and so they set up reciprocal agreements between themselves, not only to interconnect on a point-to-point basis but to re-route traffic on disaster, non-availability or congestion. In much the same way as IDD (International Direct Dial) works in the telecommunications world, ISPs work together to build an interconnected web of Internet routes.

Generally, any ISP or organisation that offers TCP/IP public data internet-working services to the general public can connect to an Internet exchange. The most prominent are the Commercial Internet Exchange (CIX) in Santa Clara, CA, USA and the London Internet Exchange (LINX) at Telehouse in London's Docklands, England. CIX has about 150 members (nearly 100 in the USA, the rest world-wide) who pay a fee to keep the exchange operational.

These exchanges can be viewed as a means to route data from the network of one exchange member to all other exchange member networks. Although there is a cost associated with connection to the exchange (to run the exchange itself) there are no settlement or traffic-based charges between the exchange members; routeing agreements are made between individual members on a peering basis. Exchanges are typically managed by an ISP or by a third party and they operate as a not-for-profit business. The exchange provides a neutral forum in which to develop consensus positions on legislative and policy issues. This role is becoming increasingly important as public and governmental interest in the Internet grows.

There is no single map or picture of all these networks, however, MDIS, one of the well-known and respected sources of Internet facts have produced an overview of the reach of the 'public' networks that span the globe (see www.mdis.com).

8.2 PUTTING THE PUZZLE TOGETHER

In this section we look at some of the practicalities of Internet connection, for instance, the variety of bandwidth options and the types of access.

Table 8.2

Usage	Bandwidth	Connection class
Dial-on-demand, single user, email, news, browsing etc.	14.4Kbps to 64Kbps	PSTN (ISDN for more bandwidth)·
Dial-on-demand, single user or system, bulk data transfer	64Kbps	ISDN
Multiple user access,[1] occasional office-hours use	64Kbps	ISDN or Private Circuit
Multiple user access,[1] frequent and 24 hour use	$n \times 64$Kbps	Private Circuit
Corporate user access,[1] frequent and 24 hour use plus operational systems communication	$n \times 64$Kbps upwards	Private Circuit, Frame Relay, SMDS
Specialist application communications	$n \times 64$Kbps upwards	Private Circuit, Frame Relay, SMDS
Simple company web site[2]	64Kbps	Private Circuit
High profile company web site[2] offering customer service, product catalogues and on-line ordering and fulfilment.	$n \times 64$Kbps upwards	Private Circuit, Frame Relay, SMDS

[1] The bandwidth requirement for multiple users depends on the number of users, their frequency of access and the nature of interaction across the link. It is often best to choose the connection class that can be enhanced to supply more bandwidth without incurring an unnecessary cost of upgrading the link.
[2] This assumes that you will want your own Internet connection from which to manage and maintain your own web presence. ISPs will be able to provide this function for you which is useful if your web presence bandwidth demands are in excess of your day-to-day access demands.

Many of the network technologies and concepts introduced already come into play here: our concern is to show how they are used to provide a well-known service. To start with, Table 8.2 gives some outline guidance on usage, bandwidth requirements and means of access.

There are several types of communications access and these can be grouped into broadly two types of access: Dynamic and Static. Each has different requirements and places differing demands on the customer of such services.

Dynamic access

In this class of access the technology used is predominantly PSTN or ISDN dial-on-demand access. Typically, this class caters for the single user, accessing the Internet from home or business. The style of interaction with the Internet and its resources is very much proactive and not reactive. What this means is that connection is used to satisfy a user's need sending email or retrieving files from the Web or from FTP servers.

During your local configuration you will need to define the local PoP that you will need to dial into. The ISP will either give you a list of phone numbers or they will be configured into the software from which you can choose. If the ISP offers a personal Web page you may be requested to supply an address, either a new DNS entry (which will incur a premium), e.g. 'yourname.yourdomain.com', or more likely, a file name, e.g. 'www. isp.com/homepages/yourname.html'.

Dynamic access in this classification also has a special meaning. When a connection is made between the customers PC/modem and the ISPs modem/router, an exchange of information between the two, at the network level, will result in the ISP network assigning the customer PC with an address for the duration of that connection. Hence this address (e.g. 194.172.36.4) is said to be 'dynamically allocated' and for this reason, dynamic access to the Internet cannot be given Internet or Domain names, of the form: name.host.domain (eg., www.bt.com).

However, dial-in users are generally given an enduring address for email. By providing an email address to a customer the ISP has to configure an entry for email routeing, known as a Mail Exchanger (MX) record within the Domain Name System (DNS). An MX record will define that any email for all the addressable users on a particular host will be forwarded to that host within a particular domain. For example mail addressed to mark@isp.com will be forwarded to the host *mailhost* within the domain *isp.com*. When the mail arrives at the host, it will be delivered to the user *mark*.

Static access

To qualify for static access you would want to connect your LAN or WAN infrastructure to the Internet for either:

- a multiple user-base, or
- for operational support systems that run real-time or batch jobs from one part of an organisation to another, or
- to communicate within a business process with a customer or supplier, or
- to maintain a permanent connection for a World Wide Web presence where 24 hour access is required for global customers as either a channel to market, a front-end to customer service operations or a shop front from which to view product catalogues, choose and order supplies and in some cases have 'soft goods' delivered on-line.

The need for Internet access is driven by many applications of communications, covering one-to-one personal and system interaction, one-to-many broadcast and many-to-many multi-cast. Static access is essential for these forms of communication where the entity with access is permanently connected and on-line 24 hours a day.

Like the dynamic examples above, static classification has a special meaning. Unlike the dynamic case, when a connection is made between the customer's network router and the ISP's network router. the customer's network becomes fixed and an addressable extension to the Internet. Because the customer is connecting another network to the ISP, rather than a single address, the customer is given (or has already pre-registered) an address range. This address range is said to be 'statically allocated' and would look something like 194.172.36.XXX. This is known as the IP address range, the example being a class C address range that can be used to address up to 255 individual devices on the customer's network, known as a sub-net. Because it has a static address, the sub-net can be configured in the Domain Name System and so becomes addressable across the Internet through a Domain name, of the form: name.host.domain (e.g. www.norwest.com). There is more on addressing in the next section.

Getting started with static connection to the Internet is less straight-forward than its dynamic counterpart as issues of naming, address allocation and security all have be handled locally.

8.3   BEHIND THE SCENES

One of the reasons for the success of the Internet is the flexibility of the protocols that lie at its core. In this section we will look at the underlying protocols and systems that make the Internet work. If you want a definitive explanation of TCP/IP and others there are many good texts that detail the ins-and-outs of IP datagram headers and exactly what ACK and SYN bits do. Some of these are listed in the References at the end of this chapter. Read on for a bare bones technical overview of Internet concepts.

The Internet Protocol

A protocol defines the expected behaviour between entities, a set of rules to which the communicating parties comply. Telecommunications and computer protocols have to be understood by machines, so they must be precise and unambiguous. The protocol needs to define the information to be conveyed, how it is sent and received, to whom it is addressed and from whom it is sent, and its meaning within its relative context.

The best known of the Internet protocols is usually referred to as TCP/IP since this is the collective term used for its many implementations (and many component protocols) on PCs and UNIX computer systems. It derives its name from two of its components IP (Internet Protocol) and

TCP (Transmission Control Protocol). In practice Internet applications use either TCP or UDP (User Datagram Protocol); the differences between these two are that UDP is for unreliable connectionless packet delivery, and TCP is for reliable connection-orientated byte-stream delivery. UDP tends to be used for simpler services that can benefit from its speed, TCP for more demanding ones where it is worth trading a little efficiency for greater reliability. The difference between the two protocols can be seen in their headers (Figures 8.6a,b). The TCP header is clearly more complex than the UDP one. Both identify the source and destination ports—the bare minimum information needed to set up an end-to-end connection. The TCP header has additional fields that allow the sequence of packets in a transaction to be tracked. Hence, any lost information can be re-

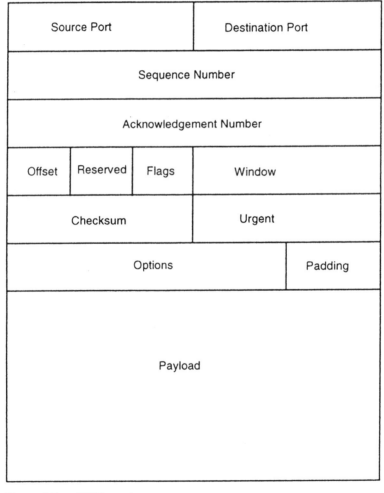

Figure 8.7a TCP header

Source Port	Destination Port
Length	Checksum
Payload	

Figure 8.7b UDP header

transmitted and both sender and receiver have some confidence that their dialogue is under control. The advantage of the minimalist approach taken with UDP is that there are few delays. So the connection may not be as well controlled but it is quick.

Communications protocols are usually portrayed as layers within a stack. This was done earlier with the seven layers of the OSI (Open Systems Interconnection) model whose concepts and layering have been retro-fitted onto a variety of proprietary communications protocol (such as IBM's SNA and Digital's DECnet). TCP/IP has not been exempt from this treatment and so the layered model is repeated (Figure 8.7).

The stacking or layering within the protocol is a convenient way to separate function in the protocol as follows:

4 The application layer defines the application software, its processes and the protocol it uses to convey its data to the communications protocol stack. In the case of email, the protocol it uses is SMTP (Simple Mail Transfer Protocol). Email applications wrap-up messages with start and end markers and attach header information about where the mail is from and to whom the mail is to be sent. This is passed to the layer below to be sent on its way; much like putting a letter in an envelope, writing an address on the front and dropping it into a postbox.

3 The transport layer wraps up the application layer message in its own data that defines the application that is sending and the application to receive; these are known as the Source and Destination ports. It will also add data to specify the overall length of the message and a number, the checksum, to use to check if any of the data it is carrying has been corrupted. This is the layer in which both TCP and UDP reside.

UDP is generally used to convey small messages of a request-response nature within single packets, such as the Remote Procedure Call we saw earlier, and TCP is used to convey larger messages within a byte-stream where some delivery assurances are needed.

The reasoning behind this difference is that, for small messages, the overhead of creating connections and ensuring reliable delivery is greater than the work of re-transmitting the entire message. To this end, TCP will attach further information to the message passed from the application to ensure that the reliable connection is maintained during the transmission and that the segments of the byte-stream all arrive at their destination.

2   The Internet layer provides the most important function of the TCP/IP stack. It structures the data into packets, known as datagrams, moves the datagrams between the Network Access layer and the Transport layer, routes the datagrams from source to destination addresses and performs any necessary fragmentation and re-assembly of datagrams. The Internet layer wraps up the transport layer data in its own data which includes the length of each datagram and the source and destination addresses (the IP addresses) which specify the network and host of the source and destination.

1   The network access layer is perhaps the least discussed of all the layers since the protocols within it are generally specific to a particular hardware technology for the delivery of data. Therefore, there are many protocols, one or more for each physical network implementation. The role of the network access layer is to ensure the correct transmission of IP datagrams across the physical medium and mapping of IP addresses to the physical addresses used by the network.

The examples above explain how an application may send data across the Internet using the TCP/IP suite of protocols, from application layer to network access layer. The reverse is also true when data is received by the network access layer. The roles of each layer are the same, the difference being that the data are 'peeled' as it ascends the stack, each layer removing the layer-specific data put there by the source TCP/IP. The concept of wrapping up data, layer by layer in this way is referred to as encapsulation.

Addressing

Each host on the Internet has a unique number, known as an IP address (this concept and the fact that there are a limited number of these were discussed earlier in this chapter). The IP address is a 32 bit number that can be used to address a specific network and host attached to the Internet. This does not mean that every Internet user has a permanent IP address. As we have seen, dial-in users using the point-to-point protocol (PPP) to connect to an ISP are 'loaned' one for the duration of their

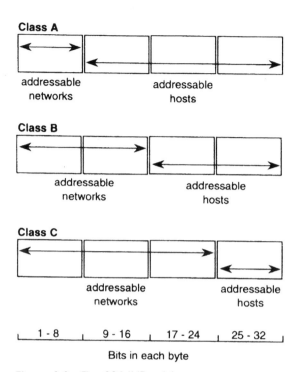

Figure 8.8 The 32 bit IP address

connection. Enduring IP addresses relate to a specific host on a specific network.

The 32 bit IP address needs therefore to be split to define both a network part and a host part (Figure 8.8). This split occurs in different positions within the number according to the class of the address, and there are four classes, A through D.

- are within the range 223 to 255.XXX.YYY.ZZZ Class A addresses are within the range 001 to 126.XXX.YYY.ZZZ - The first byte is used to define the network (bit 1 = class A, bits 2–8 the network). The remaining 24 bits are used to address the hosts on the network, therefore millions of hosts can be addressed.

- Class B addresses are within the range 128 to 191.XXX.YYY.ZZZ - The first 2 bytes are used to define the network (bits 1 & 2 = class B, bits 3-16 the network). The remaining 16 bits are used to address the hosts on the network, therefore thousands of hosts can be addressed from thousands of class B networks.

- Class C addresses are within the range 192 to 223.XXX.YYY.ZZZ - The first 3 bytes are used to define the network (bits 1, 2 & 3 = class C, bits 4-24 the network). The remaining 8 bits are used to address the hosts on the network, therefore 254 hosts can be addressed from millions of class B networks.

- Class D addresses and are special reserved addresses for multi-cast protocol address (they can be safely ignored).

There are two class A addresses missing from the list above, these are 000 and 127. These are special addresses that used for default and loopback routeing, used when configuring a host. Also, host numbers 000 and 255 are reserved; 000 define the network itself and 255 is often used to broadcast to every host on the network.

It is often necessary to define additional networks from an address range. This is done by 'robbing' host address bits for use as additional network address bits. Hence total number of addressable hosts is reduced but the number of networks increased. These networks are known as subnets. The need to define subnets is usually not technical but managerial or organisational. Subnetting caters for the delegation of address assignments to other organisations, however, subnets are only locally defined and the actual IP address of a host is still interpreted as a standard IP address.

Subnets are defined by applying a bit-mask (the subnet mask) to the IP address. If a bit is 1 in the mask it defines a network bit, if a bit is 0 in the mask it defines a host bit. It is best to define subnet masks on byte boundaries (it makes them easier to read), so a standard class A address would have a subnet mask of 255.0.0.0. All the bits in the first byte are 1 (defining the network) all the bits in bytes 2, 3 and 4 are 0, defining the millions of hosts that are addressable on a class A network.

One of the points made earlier was that IP, versatile and ubiquitous though it may be, has its limits. In particular, the range of addresses that it can support is rapidly reaching saturation point.

The response to the problem is IP version 6 (also know as IP next generation, IPNng). The main feature of this upgrade is its extended addressing range: the 4 byte field used in current IP is extended to 16 bytes, thereby giving considerably more range and variety. And that is not the only advantage of IPv6. The designers have taken the opportunity to build technology that answers a range of other questions. For instance, support is added for delay sensitive traffic (special headers for voice and video), encryption facilities are embedded and autoconfiguration for dynamic hosts is built in.

Routeing

There are two basic types of device on the Internet, either a host, such as a computer, and a gateway, such as a network router. Gateways are essential components of the Internet, without them, no two networks would be interconnected. Both these devices are required to route data both to and from hosts. The routeing decisions are not complex. When a device, such as a router, receives an IP datagram it will examine the network portion of the IP address. It will first decide that if the destination address is on the local network it will deliver the datagram directly to the host. If it is destined for a remote network it will deliver the data to the next switch en route which will make the same decisions and route accordingly.

How the router decides where to go next depends on the routeing protocol being used. The most basic option is a Distance Vector Protocol, such as RIP (Routeing Information Protocol) Routers running this type of protocol are just configured with a list of the subnetworks directly attached to that router. The router will periodically broadcast a routeing update to all other routers attached to these subnetworks, telling them about the subnetworks that can be reached. When a router receives a routeing update from a neighbour, the update will contain information about subnetworks not directly connected to the recipient router. The router will merge this newly acquired information into its own routeing table. When this router next sends its own update, it will pass on information about the networks it has just learned about. The process is repeated until eventually, all nodes have learned routes to all subnetworks.

Although effective, RIP can generate a lot of traffic and the router can spend a lot of time updating its tables. To resolve these difficulties, a new generation of routeing protocols has been created called the Link State Protocols; the TCP/IP version is known as Open Shortest Path First (OSPF). During normal working routers simply exchange 'hello' messages with their neighbours, to check that the links are working OK. These messages are short and use little bandwidth. Only if a hello handshake fails do the routers then enter a recalculation process, where all routers will recalculate the network topology.

When the network first starts up, routers will broadcast a message out of all its interfaces to try to discover its neighbours. Once a router has done this, it will broadcast a neighbour list to all routers in the network. The router is now able to calculate a complete network map from all the neighbour lists received. Only a failed link (or a new link added to the network) triggers a recalculation process.

The knowledge of what is local and what is remote is usually pre-configured into gateways by system administrators. Gateways are also configured with routeing information from more routeing protocols. These protocols pass routeing information between gateways to build routeing tables that hold information on networks and the extent of their interconnection. These routeing tables, like those above, do not hold end-to-end routes, they hold lists of which networks can be reached via their own. There are a number of these Gateway protocols: Gateway to Gateway Protocol (GGP), Exterior Gateway Protocol (EGP), (Border Gateway Protocol) and Remote Gateway Protocol (RGP) are but a few.

Mapping IP addresses to physical locations

IP datagrams are routed across networks using Internet layer protocols. When these datagrams reach their final destinations, the local networks on which the destination host resides, the network access layer protocols are able to route the datagrams directly to physical locations. The most common of these local networks are known as Ethernets.

The network access layer protocol that performs this function on Ethernet networks is known as the Address Resolution Protocol (ARP). ARP software uses a simple table which stores IP addresses and their corresponding Ethernet addresses. When a datagram, with associated IP address, is delivered to this software, ARP will look up its corresponding Ethernet address and route the datagram direct to the host. If the IP address does not have an entry in the table, it will broadcast a message to all the devices on the local network and wait for a reply. If one of the hosts on the local network receives a broadcast message containing its IP address it will respond with its Ethernet address. The ARP software will then cache the IP/Ethernet address pair for the next time that it is needed.

RARP is a variation on the ARP protocol and stands for Reverse Address Resolution Protocol. It is used by computers that do not have disks, therefore can not store any static information about network addresses. Diskless workstations can, however, request network devices for an IP address providing they have an Ethernet address. Since Ethernet peripheral devices have hard-wired addresses, the diskless workstation can use RARP to request an IP address by trading its Ethernet address.

Moving data from the network to the application

When data has been passed from your local application, down to the network through TCP/IP, routed across the network and passed to the host you have addressed, the data then has to pass back up through the TCP/IP software to the application you are communicating with. The IP protocols will ensure that the data is delivered safely to the Internet layer, but then has to determine which transport layer protocol to give it to TCP or UDP.

There are other transport layer protocols, used for passing routeing and control information. Because of this diversity of choice, each transport layer protocol has a number (these are defined in a configuration file), known as protocol numbers. TCP is 6, UDP is 17, where the number is assigned by the Internet layer, to each datagram, according to the transport layer protocol that passed it down. When the Internet layer receives a datagram from a remote host it will examine the number and pass it up to the appropriate transport layer protocol.

The transport layer faces the same problem when it needs to pass data up to the correct application process. Like transport layer protocols, applications have their own numbers, but these are called Ports (not application numbers). Well-known applications, such as SMTP (25), Telnet (23), FTP (21), NNTP (119) and HTTP (80) have well-known numbers, and like the protocols these definitions are also stored in a configuration file.

But what if an application wants to hold multiple sessions with multiple users? With a single port number, multiple users could potentially view each other's data or actions during concurrent access to a single application process. In this scenario the source port dynamically allocates a random port number to the source. This is then used by the destination port uniquely to identify the source. The combination of source and destination port numbers defines a unique network connection. In the example of simple terminal-host access, telnet uses TCP to establish a connection-orientated session between two hosts.

The combination of IP address of the source host and the dynamically assigned port number is sometimes referred to as a socket, although socket and port are often used to describe the same thing. The concepts of the socket as an addressed communications channel is widely used. A development of the basic idea is the Secure Sockets Layer, SSL, which provides secure transactions between a client and a server at the network access layer. Under SSL, the server is given a digital signature that allows a client to recognise a bona fide resource.

Naming and addressing

IP addresses are a simple and effective means of addressing individual computers on the Internet, but not as convenient as calling them names, or using a system to find a name for you if all you have is a string of four bytes separated by dots.

Using names instead of IP addresses

Internet addresses, for all their charm, are cumbersome. They are difficult to remember and easier to get wrong when typing them into an application or configuration file. Enter the Domain Name System (DNS), which allows for the friendly assignment of names to numbers in the following form:

194.172.34.25 = host.network.domain

These names and numbers are stored in a database and can be quickly and easily retrieved using a Domain Name Resolver (DNR). When a new name is created it is entered into a local DNS database. This database does not hold all the names registered on the Internet, only a few, relevant to its local domain and those remote hosts and networks most frequently accessed from the local domain. Any new name will be shared with a DNS primary name service so that all the Internet DNS information is shared beyond its local domains. Usually, a local DNS will have a primary DNS or authoritative server that it can refer to if the local DNS receives a name look-up request for a name it does not hold. The request will be passed from the local DNS, up a logical tree of DNS servers until the IP address, or name, is found. The requested information is then passed back down to the local server and cached for future use.

This tree of DNS servers can be compared with the UNIX filesystem structure and is referred to as the domain hierarchy. At the topmost level are a group of servers referred to as the root servers. Below these are organisational and geographical domains. The domain hierarchy can be visualised as an inverted tree, with a single root at the top and multiple branches stemming downwards.

Those domains below the root level are identified by using 2 and 3 letter codes, to specify either a country of origin or an organisational affiliation. Table 8.3 is not a complete list but illustrates the idea.

Table 8.3

				• (root)					
com	mil	edu	org	gov	net	uk	nl	fr	it

where

com	commercial organisations
mil	military organisations
edu	educational establishments
org	organisations (such as charities, societies etc.)
gov	governmental institutions
net	network support organisations (ISPs)
uk, nl, de	United Kingdom, Netherlands, Germany

Its interesting to note that the USA does not have a .us code to denote an organisation in the USA. They got there first (rather like the UK did for postal service) and use .com, .mil, .edu, .org, .gov and .net. However, there is a growing trend on the Internet to present yourself as a global entity and more and more commercial organisations are choosing to drop their country code in preference for a .com suffix on their domain name. There are all sort of variations occurring within local country domains, such as:

.co.uk	Commercial organisations within the UK
.ac.uk	Academic establishments within the UK
.org.uk	Non-commercial organisations in the UK
.nhs.uk	The National Health Service in the UK

The sub-division of names and numbers by the decimal point or dot does not imply any relationship. The IP address is always four bytes separated by dots. An IP address may have many (aliased) domain names associated with it. For example, two domain names 'kaa.axion.bt.co.uk' and 'trafalgar.srd.bt.co.uk' may both point to the same physical host.

As discussed earlier, to define your own domain name you will have to register with the relevant registration authority. Adding hosts to a domain, such as kaa.jungle to the bt.co.uk domain can be done by making local changes to the local DNS.

8.4 INTRANETS AND EXTRANETS

Perhaps the most concise way to define an intranet would be to say that it is the deployment of Internet technology to meet the needs of particular group or organisation. In operation it satisfies the same need as a Virtual Private Network or Enterprise Network. Because of this (and despite being built on the same underlying technology as the Internet) an intranet is quite different. The fact that it is built to the requirements of a particular set of users means that performance, security and quality of service guarantees can be designed in. The basic set-up of an Intranet is shown in Figure 8.9.

In reality, there would be numerous computers, servers and local area networks connected together. Typically, an Intranet constructed along the lines of the above figure would provide a range of information services to user terminals. News feeds, Mail, File access and on-line references would be provided along with a host of other information services. In many organisations, the Intranet provides the main source of working material, the vehicle for cooperative projects and the preferred reporting route.

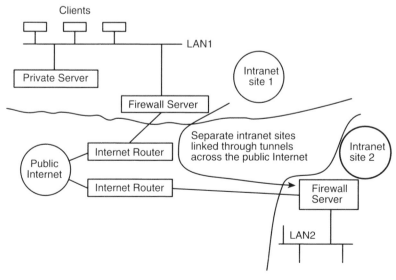

Figure 8.9 An overview of an Intranet

Because they serve a defined set of users and are maintained resources, Intranets often support quite sophisticated services. For example, a company-wide on-line directory calls might call for database access (e.g. to find an individual's details, such as a phone number and location). In instances like this, a request from a client (i.e. the browser on the user's PC) can be passed from the local server to another application via the Common Gateway Interface or CGI as illustrated in Figure 8.10. In operation, the CGI provides a script that reformats information from any system that the intranet server can see so, for instance, databases can be queried and the requested information presented to the user through a browser as if it were a sourced from the local server itself. There are several other ways in which advanced services can be provided over an intranet. For instance, mobile code (that is software in a language such as Java that can be downloaded from a server to a client and executed locally), delivered via the intranet can be used for many purposes.

Hence, end-users have access to a wide range of information—virtually anything held on a database, for instance—through a common interface. The fact that a network sits between the user and that information is, to all intents and purposes, invisible (or, at least, should be). This 'transparency of location' is one of the ideals of any distributed system that was introduced in Chapter 1. It is achieved on an intranet by a combination of the common access techniques (browser interface, associated protocols) and a wide area network that transports data between client and server as if they were co-located.

The CGI approach suits lower volumes but when company-wide scaling is required, a more robust option is needed. For instance, a directory application taking millions of hits per month would use commercial databases and application servers. In this instance, the interface to the

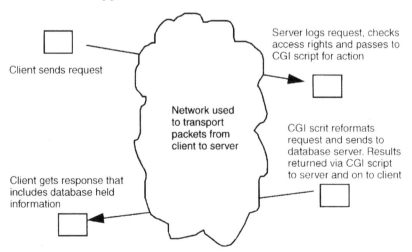

Figure 8.10 A complex transaction over the Intranet

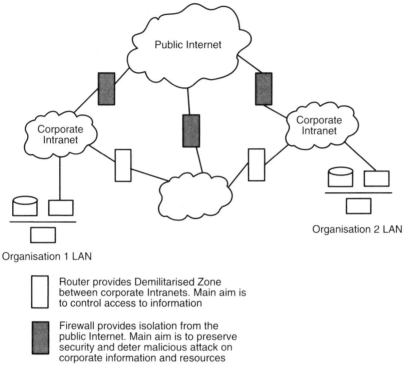

Figure 8.11 Intranets and Extranets

server would be engineered to cope with a high volume of transactions, rather than the simple protocol and script interface shown above. Some of the computing technology explained in Chapter 6 (for instance, CORBA) enables many servers to support an application and hence provide the scalability to cope with sophisticated company-wide utilities.

Intranets can be extended to form an extranet or community of interest network (COIN), that allows other organisations to connect to shared resources (directories, files etc.). When this is done, normal security concepts have to be extended slightly, as shown in Figure 8.11.

Instead of having a secure zone behind the firewall and an insecure public zone, there is a third area, referred to as a demilitarised zone, or DMZ. In general this calls for no more than a reconfiguration of firewalls, although in practice all parties need to coordinate their policies and practices.

8.5 SUMMARY

The aim of the chapter has been to illustrate the way in which a large and popular network, the Internet, works. What we have done is to take four particular perspectives: the dial-in end-user, the directly connected user, the service provider and the global view. For each of the perspectives we have examined the network options, the hardware and software components, the motives for use and the concepts that are needed for successful exploitation. Along the way we have delved into the technology behind the Internet to explain how it all fits together and what you need to acquire or know.

One thing that is very clear is that the Internet has been constructed in a very different way from the telephony network. Yet, for all of the differences, many of the same ideas and technologies are evident. The main aim in explaining some of the technical detail that makes the Internet tick, has been to foster a better understanding of those instances where the same principles are applied in a different guise and those where there is genuine divergence of approach.

REFERENCES

Naming
Albitz, Paul and Liu, Cricket (1994) *DNS and BIND in a nutshell.* O'Reilly & Associates. July.

Firewalls
Cheswick, W. and Bellovin, S. (1994) *Firewalls and Internet security: repelling the wily hacker.* Addison Wesley.

Protocols
Hunt, Craig (1994) *TCP/IP Network Administration.* O'Reilly & Associates. May.

Mail
Palme, Jacob (1995) *Electronic Mail.* Artech House Books.

Security
Oppliger, Rolf (1998) *Internet and Intranet Security.* Artech House.

General
Muller, Nathan (1999) *Desktop Encyclopedia of the Internet.* Artech House.

Intranets
Norris, Mark, Muschamp, Paul and Sim, Steve (1999) *The BT Intranet— Information by Design.* IEEE Communications, **32** No.3 59–66, March.

For definitive text on some of the technical standards used across the Internet, there is an anonymous ftp site—is.internic.net (ds.internic.net also works). The Requests for Comments, RFCs, that define IP, TCP, SMTP and all the other Internet protocols can be found here.

9

Communication Standards

The nice thing about standards is that there are so many to choose from.

Anon

If you were to apply this chapter heading to human beings and their social interaction, you might expect a few hints and tips on etiquette to follow. When it comes to less complex and disparate entities like networked systems we need to be more prescriptive. To work together effectively, there needs to be some sort of formal agreement on how systems should be built, deployed and operated. And this means that some level of standardisation is essential.

Many people spend a large part of their working lives making, developing and harmonising standards. In the broadest sense, this can entail anything from specifying the size, shape and hue of a ripe mango to defining the precise sequence of events that should be followed when two computers exchange confidential data.

The worlds of computers and telecommunications abound with standards. As will be evident later in this chapter there are a host of bodies specifying or recommending how things should be done. That is not to say that there are standards that cover all aspects of communications, nor that the standards that there are guarantee compatibility. Complexity is such that there is no communications equivalent of the Whitworth screw thread or the metric kilogram.

Given the breadth and detail of standards making, it is likely that the standards professionals are the only ones who truly understand the intricacies of the process. They are probably the only people who really know what the inter-relation of the various standards bodies is, either in theory or in practice. Even within the standards community, this is not universally known—a source of many problems in its own right. And the last thing that someone struggling with the breadth of communications technology needs is the added burden of having to understand this sort of detail.

The role of some of the main standards bodies that have something to say about communications technology is explained here, along with an outline of what their main offerings are and how they work with each other. We also present some of the more relevant deliverables from the various standards bodies—the specifications and, increasingly, products that help set a baseline for cooperation between networks and systems.

The overall coverage here is, of necessity, rather superficial. There are many issues that can only be resolved through commercial judgement, even intuition (e.g. whether to opt for the latest offering from one supplier in preference to that of another). What follows should allow some sort of positioning, though, and we at least introduce (in alphabetic order) a good number of the more prominent players who are determining the shape of today's communications technology

9.1   THE PLAYERS

We now give a brief resumé of some of the organisations that make and influence standards in communications marketplace. The list is by no means exhaustive and few of the entries are little more than introductory. Nonetheless, links are given in each case and it is relatively straightforward to get to just about any standards information you need from what follows.

ANSA

Scope
ANSA stands for Advanced Network Systems Architecture. It is a UK based industrial consortium that was established to define and demonstrate a practical architecture for distributed computer systems. It was set up under the aegis of the pre-competitive Alvey research scheme in the mid 1980's with support from UK based companies such as GPT, ICL and BT. The consortium membership was expanded to include HP, DEC and other global suppliers.

The work of the ANSA consortium was initially aimed at defining a coherent architecture for distributed processing. This is presented in the ANSA reference manual, a comprehensive and detailed guide for the designer of distributed computing systems. A second objective of ANSA was to build the software elements and test harnesses that enable the architecture to be implemented, an objective realised with ANSAware. In many ways, ANSA can be viewed as pioneering the practical application of concepts such as object orientation.

In terms of coverage, a key ANSA aim has been to bring together the three disparate areas of computing:

- desktop (i.e. the applications seen by the user)
- databases (distributed, federated and other strategies for storage)
- business systems (on-line transaction processing systems)

and to provide a common framework within which they can all be described. The most recent work carried out within the consortium has been focused on Web-based applications.

Offerings
ANSA has produced a reference manual—a very detailed technical architecture to meet its aims —along with a number of the components that are required to build real systems to that architecture (ANSAware). The results that have come out are applicable to a wide range of applications, for instance:

- general computing, such as office automation, finance packages etc.
- operational computing systems
- Web applications—the concept of Flexinet
- integration of networks and computing—relevant to Intelligent Networks and in the work of TINA.

The results have influenced other industry organisations such as the OMG, and have also been taken up by individual vendors. For instance, Telcos, such as Bell Northern, have developed their own ANSA based systems, principally for network management. ICL's DAIS was based on ANSAware.

Overall ANSA has made a strong impact on the distributed computing community. The consortium has a detailed workplan for the year ahead, although, in practice, the lifetime of the consortium is open-ended. They operate as a permanent team rather than an occasional gathering and adapt their deliverables to suit the requirements of their members.

Links to other groups
ANSA has links with other organisations, both commercial and research and standards initiatives. Specifically:

- The Object Management Group—ANSA has direct participation and is generally regarded as providing much of the architectural input. It also influences the input of the individual vendor members;

- TINA-C—the work here focuses on telecommunications and can be regarded as complementary;

- ISO Open Distributed Processing—ANSA have made direct contributions to the work and have also influenced the input of member companies (both nationally and internationally).

Overall, much of the ANSA output has been adopted by other standards bodies. This is not really surprising as the initial task within ANSA was to harvest seminal work by research institutes.

    http://www.ansa.co.uk

EIA

Scope
The Electronic Industries Association is a US based organisation of electronic equipment manufacturers. As the name suggest, the main focus of the group is on standards for physical devices such as computers and their peripherals, although they also concern themselves with more abstract issues such as the exchange of information between modelling tools.

Offerings
Many of the EIA standards have achieved global acceptance. Perhaps the best-known and relevant standards are those for the connection of Personal Computers. The RS-232 interface standard is universally used to connect PCs and peripheral devices.

    One of the more recent EIA standards that has significant bearing on communications is their Component Definition Interchange Format (CDIF). This defines how information is exchanged between modelling tools, and between repositories, and defines the interfaces of the components to implement this architecture.

Links to other groups
The EIA coordinates its activities with other global standards makers such as the ITU-T. Indeed the RS-232 standard is directly compatible with the ITU-T V24 standard. The EIA also work with the Object Management Group on CDIF (in particular on how the meta-model of the Unified Modelling Language (UML) can be used with the CDIF Transfer Format or the CORBA bindings to interchange object analysis and design models in a tool-independent manner). As a result of this, CDIF is being incorporated into the OMG's XMI standard.

    http://www.eia.org/

ETSI

Scope
The European Telecommunications Standards Institute is an open forum that develops standards across the whole breadth of technology. ETSI

consists of several hundred member organisations (both operators and manufacturers) drawn from nearly 50 countries. The remit of ETSI is to develop voluntary standards to improve pan-European telecommunications services.

Offerings
ETSI produce a wide range of standards. Perhaps the best known is the digital cellular mobile standard, GSM.

Links to other groups
The ETSI programme of work is coordinated with that of the ITU-T.
    http://www.etsi.org/

IAB—Internet Architecture Board

Scope
The Internet Activities Board (IAB) is an influential panel that guides the technical standards adopted over the Internet. Linked to the IAB are two bodies that do much of the down to earth work to make the Internet a practical reality. The better known of two is the Internet Engineering Task Force (IETF), the source of the operational and technical standards on the net (captured in the widely used Requests for Comment, RFCs). The Internet Research Task Force makes up the trinity. This body takes on the research questions for future development of the Internet.

A more recent extension to these long-established bodies is the Internet Society, which aims to provide a forum for disseminating information and discussing future directions. Internet information is also available from the Internet network information centre (InterNIC), a central repository for everything from how to connect through to directory services.

Offerings
The IAB is responsible for the widely accepted TCP/IP family of protocols.

They have also a raft of protocols that were initially used in the Internet but have subsequently been more widely adopted. Examples include the network management protocol SNMP, the mail protocol SMTP and the Hypertext Transfer Protocol HTTP.

The most widely used of the IAB offerings are probably the RFCs, which define all of the various Internet standards.

Links to other groups
The IAB is interesting in that it is open to virtually anyone who wishes to contribute. It has traditionally comprised a mix of researchers, networkers and computer developers. Given its fairly pragmatic approach it has tended to work in parallel with the more formal standards bodies. It does

link, though, mainly in mode of Internet standards being offered for adoption by the likes of ISO.

http://www.iab.org/
http://www.isoc.org/

IEEE

Scope
The Institute of Electrical and Electronic Engineers is one of the largest professional bodies in the world. In addition to its other professional activities, it has also produced a number of important standards that have been widely adopted.

Offerings
The main standards offerings of the IEEE are in the area of Local Area Networking. Their 802.X series describe all of the various types of LAN in use (e.g. IEEE 802.2 is the standard for an Ethernet LAN)

Links to other groups
Some of the IEEE standards have been adopted by ISO. For instance, the IEEE 802.2 specification is the same as ISO 8002.

http://www.ieee.org/

ISO

Scope
ISO is an international organisation responsible for standardisation in Information Technology. The name is commonly believed to stand for International Standards Organisation. In fact ISO is not an abbreviation at all. It is intended to signify commonality (the name deriving from the Greek Iso, meaning 'the same'). The main focus of ISO is on data communications and protocols. It is allied to the International Electrotechnical Committee (IEC), who are concerned with connectors and environmental aspects of technology.

Offerings
ISO is responsible for many data communications standards but is probably best known for its seven-layer Open Systems Interconnection (OSI) model. This model defines the distinct logical layers in a communication systems: Physical, Data Link, Network, Transport, Session, Presentation and Application. As well as giving an overall framework for peer to peer interaction, ISO specifies many of the options for what goes inside each of the model's layers as well as what goes between them (i.e. the peer to peer protocols that link the logical layers).

Links to other groups
Many organisations feed into ISO.
  http://www.iso.ch/

ITU-T

Scope
The International Telecommunications Union is a specialised agency within the United Nations and is responsible, as the name suggests, for Telecommunications. There are nearly 200 ITU members across the world and the activities divide into two main areas, the telecommunications area, under the auspices of the ITU-T and the radio area, which falls under the ITU-R.

The ITU-T was previously known as the CCITT (Comité Consultatif International de Telegraphique et Telephonique). The ITU-T (and the CCITT before it) publishes recommendations. These are, in truth, firm standards for public telephone networks. The ITU-T cover all aspects of telecommunications from planning through to operation and from basic telephony to advanced data services

Offerings
The output of the ITU-T is a set of published recommendations. These appear at regular intervals in the form of a structured set of standards. They tend to be referred to by terms like 'red book', which indicate the currency of that set of recommendations. ITU-T output follows a set structure with each category allocated a specific letter of the alphabet, for instance:

  G-series for transmission standards (such as G703)
  I-series for ISDN standards (such as I430)
  Q-series for switching and signalling standards (such as Q3)
  X-series for data networks standards (such as X21)

The coverage in each of these areas is vast. There are a huge number of established ITU standards in just about every conceivable area of communications.

Links to other groups
The ITU is linked to all of the other organisations.
  http://www.itu.int/

OAG

Scope
The Open Applications Group (OAG) is a non-profit making consortium of application software vendors that aims to create common standards for

the integration of enterprise business applications. Work includes integration between enterprise planning and managing systems, integration to other business applications and integration to execution systems.

Offerings

Key standards from OAG include: OAGIS—Open Applications Group Integration Specification and OAMAS—Open Applications Group Middleware API Specification The Open Applications Group's Integration Specification (OAGIS) is targeted at integrating the business objects which contain the main business functions that occur within an enterprise.

The OAG has also created an architecture, termed the Business Object Document (BOD), to provide business object integration across heterogeneous environments, including multiple hardware platforms, operating systems, databases, middleware etc.

Middleware API specification (OAMAS) is a proposal for a common way of connecting business applications to each other at the technical level. It is not built on a specific architecture or middleware but, like OAGIS is 'Technology sensitive, but not technology specific'.

Links to other groups

OAG is most closely linked to the Object Management Group (OMG) and the Open Group. Their work complements that of both these bodies.

http://www.openapplications.org

Object Management Group

Scope

Founded in May 1989 by eight companies: 3Com Corporation, American Airlines, Canon, Inc., Data General, Hewlett-Packard, Philips Telecommunications NV, Sun Microsystems and Unisys Corporation as a non-profit corporation it now includes over 800 members. OMG is based in the USA, with marketing partners in the UK, Germany, Japan, India and Australia. It was set up to promote standards and guidelines for developing and managing distributed objects and, more recently, component-based developments.

The OMG's main aim is to develop an Object Management Architecture (OMA) and establish CORBA as the 'Middleware that's Everywhere'. OMG standards are open standards that can be implemented by vendors in a variety of ways. The OMG operates a number of Task Forces and Special Interest Groups to develop standards for domain-specific interfaces including:

- Business Objects
- Manufacturing

- Electronic Commerce
- Telecommunications
- Financial
- Medical

OMG specifications are chosen from submissions by member companies, rather than being developed wholly through the committee processes used in ISO and ITU-T. As such, OMG specifications are *de facto* rather than *de jure* standards.

Offerings
Object Management Architecture (OMA) is set of standards produced by the Object Management Group (OMG) to create a component-based software marketplace by encouraging the introduction of standardised object software. Particularly, it is aimed at establishing CORBA as the 'Middleware that's Everywhere'. Some of the key standards include:

- CORBA
- IIOP (Internet Inter-ORB Protocol)
- IDL (Interface Definition Language)
- UML
- Meta Object Facility (MOF)
- Transaction Processing Standard
- XMI

As well as stimulating interest in the market-place for object-orientated technologies, OMG also produces specifications for product procurement.

Links to other groups
The OMG has ties with many organisations, most notably with the Open Group, who take responsibility for handling conformance to OMG specifications and have also identified OMG specifications within their own work.

ANSA are directly involved in the OMG and also influence the input of its member companies. The OMG has identified the ISO Open Distributed Processing standardisation initiative as being of strategic architectural importance and requires statements of architectural conformance to ODP for all submissions (although it is not necessarily true that OMG specifications will actually be conformant).

It may seem that the OMG overlaps with the ISO Open Distributed Processing initiative. In practice, though, the OMG's considerably shorter

time scales mean that the activities are more complementary. It is anticipated that ODP will provide the architecture and OMG the detailed
component specifications.

  http://www.omg.org

Open Group

Scope
The Open Group is an international consortium of vendors and customers
from industry, government and academia. It defines standards to support
corporate IT users. Its IT DialTone initiative infrastructure aims to help
organisations to evolve their computing architectures to meet the needs of
participating in a global information infrastructure.

  There are two well-known organisations that joined to form the Open
Group—the Open Software Foundation and the X/Open group. The
former developed the Distributed Computing Environment (DCE), recognised as the strategic direction for distributed computing by many.

Offerings
The Open Group aims to deliver software that is subsequently incorporated into commercial products. There is typically something like a 6-
month lag between the release of software by the group and its re-
emergence in a product from a major vendor.

  On the specification side, one of the key standards is the Open Group
Architectural Framework (TOGAF). This is a tool to help define an architecture for an information system. It is based on the Technical Architecture
Framework for Information Management (TAFIM), developed by the US
Department of Defense. TOGAF is not an architecture itself, nor does it
define an architecture. It provides guidelines for developing specific
architectures through a set of services, standards, design concepts,
components and configurations.

Links to other groups
The Open Group has close working relations with the OMG and Telecommunications Management Forum (TMF).

  http://www.opengroup.org

SPIRIT

Scope
Service Providers Integrated Requirements for Information Technology
(SPIRIT) is an initiative that came from the Telecommunications Management Forum. It is a consortium of telcos, aided by their major IT suppliers,
jointly specifying a general purpose computing platform. Rather than

producing standards, SPIRIT's aim is to provide guidelines that help a buyer to select appropriate system components.

Offerings
The benefits from SPIRIT follow on from those associated with the deployment of open systems. The fact of the matter (at least for the time being) is that the open systems movement has resulted in an enormous diversity of both standards and products. Whilst increasing choice and reducing costs through greater competition, any potential benefits that may have been attained are offset from the additional cost of supporting this diversity and coping with the irritating interworking issues that come with it. As one salesman of proprietary computing equipment was heard to say: 'We don't want to get locked into open systems'!

SPIRIT is all about the specification of open systems platform components, the aims being to reduce the diversity of choice, to increase the chances of inter-operability and portability and to decrease the breadth of skills required to develop and maintain such systems.

In effect, SPIRIT shares the cost of many organisation's in-house activities and provides a partial (if not full) replacement of the company specific guide. It also provides a coordinated statement to the IT supply industry from a large industry sector (i.e. telecommunications).

Links to other groups
SPIRIT is 'harvesting' activity that has selected the most pragmatic standards from the most appropriate standards body or vendor.

http://www.tmforum.org

TINA-C (Telecommunications Information Networking Architecture-Consortium)

Scope
TINA is a telecommunications domain-specific open software architecture developed by a consortium (TINA-C) of over 40 of the world's leading network operators, telecommunications equipment and computer equipment manufacturers. It works to integrate World Wide Web, multimedia and current computer technologies with the more traditional and mature telecommunications technologies. TINA states it main goals as:

- Make it possible to provide versatile multimedia and information services

- Make it easy to create new services and to manage services and networks

- Create an open telecommunications and information software component marketplace

The common architecture around which new services can be built and deployed is derived from an object-orientated based analysis and is implemented by software components in a distributed processing environment. There is a strong decoupling between the applications themselves and the distributed processing environment (DPE) which, although derived from objected-orientated analysis, is not necessarily OO in implementation.

Offerings
TINA-C is concerned with pre-competitive research and as such the benefits lie more in better understanding and shared views than common products. The main output of the group is their architecture. This is a four-layer model comprising:

- Hardware layer—processors, memory, communication devices.
- Software layer—operating systems, communications, and other support software.
- DPE layer—provides support for distributed execution of telecommunications applications.
- Telecommunications Applications layer—implements the capabilities provided by the system.

TINA is further divided into four sub-architectures.

- Computing Architecture—concepts and the CORBA based DPE.
- Service Architecture—principles for providing services.
- Network Architecture—model managing telecommunication networks.
- Management Architecture—principles for managing software systems.

Links to other groups
TINA-C draws on the efforts in ODP and OSI, and has direct input from ANSA.
   http://www.tinac.com/

Telecommunications Management Forum

Scope
The work of the Telecommunications Management Forum, TMF (formerly known as the NMF—Network Management Forum) covers all of the operational aspects of running a telecommunications network or service. The initial focus of the group was on the interfaces and procedures needed

to manage complex networks comprised of many different elements. The fact that many of the components of modern networks are computers has broadened the Forum's remit.

The service based approach of the TMF to the management of networks and systems covers the traditional concerns of both telecommunications networks and distributed computer systems.

There are over 100 members of the TMF and there is a fairly even balance between all of the major trading regions and across the computing and telecommunications industry. The Board members are, AT&T, BT, Bull, DEC, HP, IBM & STET (Italy).

Offerings
The TM Forum is the only industry consortium that is interested in covering the full scope of inter-operable management systems. This means everything from requirements through to product delivery and testing. Results of note have been the delivery of:

- The OMNI*Point*™ specifications for management systems. These introduced the CMIS and CMIP standards for interfacing network elements and management systems.

- SPIRIT—an initiative to capture service provider requirements for Computing Platforms (jointly with the Open Group).

- AIMS a group of smaller vendors who have developed the Open Edge —a legacy system integration and migration technique.

- TOM & TIM—the Telecommunications Operations Map and the Telecommunications Integration Map. These two documents present a structure to support all of the operational functions of a network and principles for the deployment of systems to support them

The Forum holds a detailed workplan proposal that covers a two-year period. There is also a major conference (Telemanagement World) every 6 months to report on developments and to seek member input on where effort should be focused in the future.

Links to other groups
The TMF has positioned itself as the Consortia of consortia for all matters concerned with Service and Network Management. It has pro-actively sought to cooperate with all other interested groups who in the main have management on only a part of their agenda. This includes, for instance, the various broadband technology groups (the ATM & Frame Relay Fora etc.) as well as the groups mentioned above.

In the main, the Forum overlaps with just about all of the other organisations which carry out some work on service and network management

but does aim to ensure that work programmes are aligned and that duplications are kept to a minimum.
http://www.tmforum.org

There are numerous other standards bodies that .could have been included here—there always will be. A few of those likely to have some impact on future communications standards are:

DAVIC, the Digital Audio Visual Council is rapidly setting out how networked multimedia should be handled.
http://www.davic.com

The World Wide Web Consortium (W3C) which promotes standards for the World Wide Web and guide its evolution to achieve its full potential.
http://www.w3.org

ETIS, the European Telecommunications Informatics Service which works with all of the other European bodies, such as ETSI, to develop and promote networking-orientated standards
http://www.etis.org

And then there all of the national standardisation bodies such as DIN in Germany and BSI in the UK, the major European research bodies such as Eurescom, the network technology for a (e.g. ATM forum, Frame Relay forum) and influential Government organisations such as the US DoD. All have their part to play and all contribute in some way.

We could go on, but that would be to defeat our purpose of giving a brief tour. As stated at the beginning, understanding who the standards bodies are and what they are doing is only one part of the picture: politics, alliances and user preference all play their part in determining what is used in reality. In the light of this limitation, we will now move on from looking at the visible players in the standards arena.

One final point in this section is an observation. This is that it has frequently been the less formal bodies that make the biggest impact in computing (e.g. ANSA and the IAB) but the more formal, consensus orientated bodies (the ITU and ISO) who dominate with telecommunications standards. The main reason for this is the traditional contrast between the fast moving and competitive nature of the computing market and the more considered and integrated world of telecommunications. Things may change as the two come together, though, and this pressure is being increasingly applied by competition and a need for differentiation.

9.2   HOW THEY WORK TOGETHER

If standards are to promote interworking there needs to be some sort of overall master plan that join the efforts of the various players. Most of the organisations mentioned here do have a declared strategic relationship with one or more of the others, so there is some coordination (albeit fairly loose). In general, though, there is a pattern of cooperation that ensures a consistent outcome.

Some of the players mentioned here (e.g. the TMF) can be regarded as harvesting. They select standards and tailor the output of other standards bodies to produce integrated solutions that can be applied by an end-user. Others (such as the OMG) are more concerned with delivering technology. They produce component specifications and technologies in the form of reference implementations. And, of course, others still, such as ITU-T and ISO, produce the definitive recommendations that most people would recognise as standards.

This may all seem convoluted and forbiddingly complex (at least at first sight) but, once seen in operation, it is all quite rational. Figure 9.1 illustrates the flow of some of the communications solutions from their concept and specification (mostly the remit of formal standards) through to delivery of product (the prime aim of the industry consortia). From the end-users point of view, there is a spectrum of standardised offerings from formal standards (useful for reference) through usable selections of standards (that relate to a specific area, useful for design) to conformant software components (essential for systems building).

And so the above figure gives the general picture of cooperation between the various standards bodies, even if it is tacit cooperation. To understand how this actually translates into working practice, we can take an established example: the OMG's Common Object Request Broker Architecture (which also illustrates the potential influence gained through collaborative research).

The OMG specifications, were submitted by vendors based on their products, and were influenced by the results of ANSA and also ISO's Open Distributed Processing initiative. As well as being implemented directly by vendors (conformant products are available) the CORBA specifications were adopted as part of the Open Group's DCE. The Open Group also

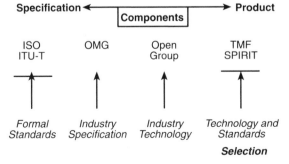

Figure 9.1 The standards 'life-cycle'

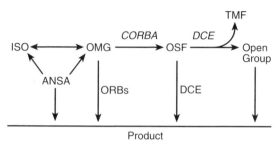

Figure 9.2 An example of the flow of ideas through the standards bodies

endorse the CORBA specifications in their own right. This flow is illustrated in Figure 9.2.

Finally, one small caveat. The fact that industrial consortia play a major part in determining many of the standards means that commercial pressures inevitably play a major part in what happens. There will inevitably be some diversity in the marketplace and some value judgements to be made by the users as to the best product for their needs. This contrasts with those more regulated areas, such as numbering or addressing, where the standards are much more technical agreements that everyone expects adherence to.

9.3 SUMMARY

To many of us, standards just happen and have to be suitably acknowledged, adhered to or coped with, rather like the weather. Often they creep up on you and are quietly accepted. Occasionally, they come like a bolt out of the blue and cause a major headache. Some groups of people even have the motivation, or wherewithal, to make the weather.

A little awareness of what these people do and how they do it can be valuable. And that is what this chapter has covered. There are a number of points that relate to communications standards:

- The first is that there is no one body to provide everything you need. This is not really surprising when you consider the diversity and complexity of the area. Some of the main standards bodies have been described here.

- Second, the standards that have been produced are not uniform in nature: some are technical (protocol definitions, architectural specifications), some are more orientated towards supply, management or operations.

- Third, not all standards are badged as 'standards'. Some are called recommendations and some are produced by industry groups, rather than by official standards bodies. The best bet is to regard any widely held agreement as a standard.

- Finally, it should not be forgotten that standards are but one piece of the jigsaw. Market forces, vested interest and pure chance sometimes prevail over all manner of consensus

This chapter has explained where the drive for standards has come from and who the main players are. The intent has been to give an appreciation of why the current situation exists and to impart some understanding of how it is likely to evolve—the motivators and drivers for change.

One thing that should be clear is that the standardisation of communication equipment, protocols and components (or anything else, for that matter) is a complicated affair. Nonetheless, it is vital to effective integration and interworking of systems and network—all cooperation relies on mutual understanding. For all the complexity that is involved, some form of negotiated consensus is needed in a diverse world. To quote Darryl F. Zanuck, 'If two men on the same job agree all the time, then one is useless. If they disagree all the time, then both are useless'. The latter scenario is most definitely one to avoid!

10

A Meeting of Minds

You can't depend on your eyes when your imagination is out of focus

Mark Twain

In spite of their separate heritage, it should be quite clear that computing and telecommunications technologies are on a common course. The emphasis, approach and language may be different, but both communities are intent on allowing people to communicate—anytime, anyhow, from anywhere. In this chapter, we look back at the major trends identified at the start of the book to see how each of them is being realised by one or more of the latest advances in technology.

Of course, this retrospective approach does not answer the question that interests most people: what happens next. The truth is that no-one really knows the answer, and it is often those closest to the subject who make the poorest guess. In view of the perils of prediction, the main part of this chapter explores a scenario that has been widely aired as the future of communications, that of an IP based network that carries all voice and all data traffic.

One thing that is clear as we fit together trends in communications and the technologies that drive them is the level of convergence between all of them. For instance the ability of a PC user to hit an enquiry button on a screen which connects them to a call centre draws on a host of technical capabilities. It is the assembly and presentation of these technical capabilities that counts.

10.1   ARE WE HAVING FUN YET?

If we look back at our five main trends for communications, it is clear that there are technology advances that promote each one of them. In fact, it is interesting to note just how neatly each of the major initiatives that adorn the telecommunications and computing trade press impact on one or more of them, thus lending credence to their coming to pass.

From Narrowband to Broadband
The speed with which the Integrated Services Digital Network and data services such as Frame Relay and ATM are now being taken up reinforces this trend. Just over the horizon, there are more initiatives aimed at getting more bits to information hungry consumers. The use of Wavelength Division Multiplexing (WDM) in optic cables will increase transmission capacity and this will be delivered to the user by xDSL technology.

From Voice to Multiservice
It is instructive to look at the products on offer from the Telcos. Until very recently, the range was limited to telephony and related services. Over the last few years there has been an explosion with a vast array of voice, data and information services now available. One of the opportunities that a high capacity network offers is greater flexibility in the services that can be provided.

From Wires to Wireless
One of the most dramatic success stories in any industry would have to be the growth of the mobile telephone business. The advent of the Universal Mobile Telecommunication Service (UMTS) will enhance the already impressive set of services that multi-mode mobile terminals (e.g. the Nokia communicator) provide, with location identification and a raft of other communication services.

   And with the Bluetooth initiative defining a short-wave radio connection standard that will enable devices such as a TV, HiFi, Video Recorder, Games Terminal and Computer to link together, the prospect of totally networked world looks close at hand.

From Desktop to Core Network
We described the technical elements of the Internet in Chapter 8. There are features on its inexorable rise in just about every technical journal and graphs showing that the whole world will be on-line by 2010 are hard to avoid. Less heralded is the fact that the traditional uses of the Web are expanding about as fast as its user base. As well as an information resource, it is now being considered as a viable alternative to the telephone network—voice over IP (VoIP) is a serious contender for a considerable chunk of the telephony market. Incidentally, VoIP is not a new concept, it was first put forward in the 1970s in RFC742. Twenty years on, major

software vendors such as IBM and Microsoft have bundled voice enabled IP software into their operating systems.

The technology that enables a best-efforts, connectionless network carry acceptable voice, with its sensitivity to delay, is demonstrable. Issues such as quality of service and service level assurance have yet to be resolved but this is no surprise—telephone engineers spent years building and honing the techniques to deliver a quality product. Also it seems reasonable that the ideas covered in Chapter 3 will endure, as the issues that need to be tackled are the same, even if the technology used is different. In section 10.3 we illustrate the point a little more by considering what a world based on IP would look like—and what sort of challenges it would pose.

From Switched to Intelligent Networks
As intelligence has been added into the telephone network, it has started to look more and more like a computer network. Indeed, the next generation of telephony switches will be multi-functional processors loaded with sophisticated control software. The question then arises 'where are the Application Programming Interfaces (API) on these computers?' If public networks could present the same sort of open interfaces as the computers they are built from, then this could lead to an explosion in the development of customised software applications. And this could fire the communications industry in much the same way as the now ubiquitous DOS operating system led to a plethora of programs for personal computers. Just how this may come to pass is described in the next section.

10.2   IT'S GOOD TO TALK

Traditionally, public network operators have been extremely guarded about allowing access to the intelligence that controls their networks. The reasons for this are fairly obvious: they want to preserve network security and counter the threat of unauthorised tampering and, even if access is legitimate, they want to safeguard network integrity so that individual applications cannot cause widespread damage. It is this desire for control over the configuration of the network that has driven much of the historical development in telecommunications networks.

It has been a very different story for the suppliers of computer technology. A diametrically opposite view has been taken across the industry, with open systems being a shared goal—witness the number standards bodies referenced in Chapter 9 with the word 'open' in their title. The tendency has been towards 'plug and play' computing components with issues of security and feature interaction left for the user to take care. The plummeting price of computer systems and huge number of suppliers is testament to the success of the open approach.

But now, one of the key psychological barriers between the telecommunications and computing worlds is set to fall. An API, or Application Programming Interface, has been defined for telecommunications service. This API, known as Parlay, allows anyone to develop an application that can be 'run' on the network, in much the same way that a word processing package runs in a windows environment. Hence, the imagination of many software developers can be on tailored solutions to meet individual telecommunications service needs.

The closest historical parallel would be what happened with the PC after the creation of the Disk Operating System (DOS). Because anyone could write an application to run on DOS, small niche market players generated thousands of software packages. And this all happened in a fraction of the time that it would have taken the computer manufacturers to write the same applications. If the communications industry follows the same path, the future will be all about thousands of niche markets rather than the small number of big ones we now recognise.

There is every reason to expect this to happen. The development of the API is in the hands of an industry-working group comprising key players such as Microsoft, BT, Nortel Networks and Siemens. Applications could include personal routeing information to enable calls to be transferred to voicemail, email or a cell phone under the control of an application running on a customer's Personal Digital Assistant rather than by a centralised network service. This more intimate integration of intelligence at the edges of the network and its core would lead to an 'explosion' in the number of customised services being created.

Built around a network-independent framework interface, the API allows network operators to define and provide controlled access to network resources by third parties. It includes authentication and billing capabilities to guard against malicious or unbilled use of network resources and contains specific instructions on how functions, such as call routeing, voicemail and email, operate within a network.

The Parlay API has been designed to work with existing service control interfaces such as the Intelligent Network protocols described in Chapter 7 or Microsoft's Telephony Application Programming Interface (TAPI), outlined in Chapter 1. Given this, voicemail services and call centres using Computer Telephony Integration (CTI) would be early adopters of the technology. But the potential benefits of being able to customise applications without the full technical and operational involvement of the telecommunications operators would change the whole industry.

10.3 TWO BECOME ONE

A simple proposition—if telephony is a 64 kbps service and data communications capacity is growing at such a pace that it can readily support

millions of 64 kbps connections, why is there no single network for all forms of communication? Let us test this idea—let us imagine a network that can be built and dimensioned to carry all of the domestic voice and data traffic in a single industrially developed country.

It would be quite reasonable to assume that this network would look rather like the Internet. After all, here is a network that has grown at consistently rapid rate and can demonstrably support many forms of communication. There is a popular belief that the Internet works on a different time scale from the established telecommunications world with 1 Internet year = 7 Telecommunication years. So this already looks a good horse to back.

Things look even better when we look at where communications are going. Over the last few years data traffic has been growing at approximately 20% per annum compared with only 8% for voice. So having our universal network strong on data communications seems a 'Good Thing'.

And for a final sanity check, let us consider technology. We already know that the technical mainstay of the Internet—the Internet Protocol—provides a common standard for virtually all networked applications. But what about the capacity needed to deal with the enormous traffic levels that the PSTN can deal with (e.g. 26 million calls a day in the UK alone). Even here, we seem to have a good story. The basic building block of the Internet is the router, and the latest generation of these devices is capable of switching Terabits of data per second. It seems that we may have enough vroom for the job!

All looks well. And many share our imaginings, as illustrated in Figure 10.1 with the predicted revenues from IP based telephony. Before dashing

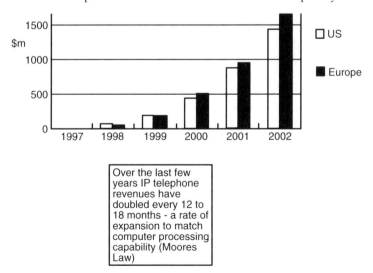

Over the last few years IP telephone revenues have doubled every 12 to 18 months - a rate of expansion to match computer processing capability (Moores Law)

Figure 10.1 A view on IP Telephony growth

off to install our universal communications fabric, we should dwell on a few practical issues. For real-time voice communications, something that we have taken for granted for the past 20 years, is a good quality of service, reliable connection and well managed network. But with IP, data is transported in packet format so it is inherently non-deterministic. So, for any delay sensitive traffic to be sent on an IP network some form of Quality of Service (QoS) assurances within the network are required. And this implies that the currently available technology needs to be extended.

Mechanisms to implement some form of Quality of Service in IP networks have been under investigation for several years. The most promising mechanism for implementing QoS is for the customer to set the Type of Service (TOS) bits in the IP header to indicate what QoS the particular packet requires. The TOS bits can be set to indicate that the information is either real-time (e.g. VOIP) or requires some predetermined QoS. The TOS information is used by the access router to first ensure that the customer's data is within the agreed contract and then to place information with similar QoS requirements in similar output queues. The basic idea is that high priority/low latency information is given the highest priority and hence is offered first access to available transmission. One technique that is currently being standardised is MultiProtocol Label Switching (MPLS) which is based on the proprietary Tag implementation from Cisco.

With MPLS, a number of label switching routes are set up between the various routers in a network with each router having a label forwarding table. On ingress to an MPLS network the destination address of the incoming packet is evaluated and a label added to the packet that indicates the next router and the priority of the packet. All packets to the same next router and with the same QoS requirements are assigned the same label. The output queue of the ingress router and the links between routers are engineered so that set amounts of bandwidth are allocated to certain labels and hence QoS type packets. By this mechanism the packet travels from router to router with its label being swapped but always going via a traffic engineered route for its particular QoS requirements. At the egress router the last label is stripped off and the packet is transmitted to its destination. (see Figure 10.2.) As well as supporting voice traffic, this technique can be used to separate traffic on different virtual networks: different label routes do not interfere with each other, so the network can be segmented by usage.

It would seem that we have all we need. If the technology progresses as it has done in the past, then there is every prospect of components that have the power to cope with volume telephony as one part of a the multimedia traffic load. But before we move on, it is worth casting our minds back to some of the issues raised in Chapters 2 and 3. If delay sensitive traffic is given priority then it looks like our IP based network is being forced into setting up connections. These may not be real connections (in the telecommunications sense) but we still have to consider whether we

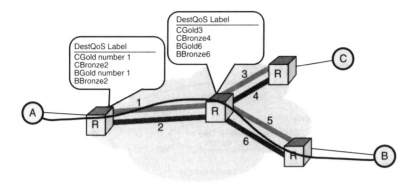

DestQoS Label
CGold3
CBronze4
BGold6
BBronze6

DestQoS Label
CGold number 1
CBronze2
BGold number 1
BBronze2

Each packet carries an indicator to show what level of service it expects. At each link in the end to end connection, the packet is routed to the appropriate path, in this case, the gold path (which has some pre-specified assurances over delay etc.)

Figure 10.2 From A to B with MPLS

are introducing blocking in network type that has not had to deal with the problem before. And, with network paths being pre-ordained, some concept of call admission control needs to be introduced, so that alternative routes can be taken when specified link is up to its capacity.

One final thought is that the management of our universal network might be tricky. With routeing decisions taken on a link by link basis, there is no real view of where the end-to-end path is. So, if a router were to break it would be difficult to see which phone calls it affected (if any, as the packets might simply divert). Also, if part of the network is segmented to provide a Virtual Private Network, correlation between network events/status and user services would have to be made, and this is something that is yet to be sorted out.

So, for all of its appeal, our vision does have some blind spots. Only time will tell if these clear.

10.4 SUMMARY

In this chapter we have considered the 'disappearance of telecoms'. We have imagined a world where all communication takes place over a souped-up Internet. And, despite there being some unresolved issues, it does not look too far-fetched an idea.

Much of this book has been about existing technology. In this chapter we have looked a little over the horizon but this has been more at how we use what we have rather than any predictions on new technology. In doing this there is one thing that does come to the fore and this is our

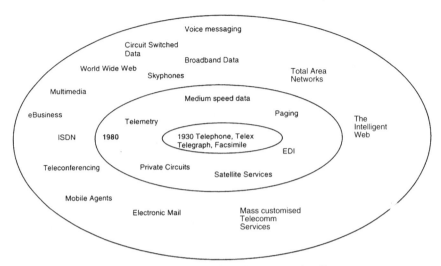

Figure 10.3   The ever expanding realm of communications

reliance on communications. Instant, perfect and rich communication is
the norm these days. People expect to be able to send and receive just
about anything to anyone they please from wherever they are.

Communications is now so embedded in society that it can be thought
of as logical underwear—something that is rarely noticed unless it is
absent. And, who knows, with the advent of wearable computers, the
logical and physical might merge into one.

REFERENCES

Chatterjee, Samir (1997) *Requirements for Success in Gigabit Networking.*
   Communications of the ACM, July, **40,** (7), 64–73.
Atkins, John and Norris, Mark (1998) *Total Area Networking.* John Wiley &
   Sons.
Tatipamula, Mallikarjun and Khasnabish, Bhumip (1998) *Multimedia
   Communication Networks.* Artech House.
Whyte, Bill (1999) *Networked Futures.* John Wiley & Sons.

Glossary

Tyranny begins with a corruption of the language

W H Auden

The convergence of computing and telecommunications has led to a broad and often confusing set of terms that are assumed by many in the profession to be widely known. In truth, even the professionals often misunderstand each other.

Here we list here many of the key abbreviations and concepts in common use. Some of the terms have been explained in the main text, many are not. In either case the aim is to clarify some of the more complex ideas by giving some idea of their context, application and relationship to one another.

A

Abstraction	A representation of something that contains less information than the something. For example, a data abstraction provides information about some reference in the outside world without indicating how the data is represented in the computer.
Access control method	A methodology of distinguishing between the different LAN tech nologies. By regulating each workstation's physical access to the transmission medium, it directs traffic round the network and determines the order in which nodes gain access so that each user obtains an efficient service. Access methods include Token Ring, Arcnet, FDDI and Carrier Sense Multiple Access with Collision Detection (CSMA/CD), a system employed by Ethernet.
Active-X	Microsoft technology for embedding information objects and application components within one another. For example, an Active-X button can be embedded in an HTML page that is displayed in a browser window.
Address	A common term used both in computers, telecommunications and data communication to designate the destination or origination of data or terminal equipment in the transmission of information. Types

of address include hardware addresses (e.g. 0321.6B11.3310, for an Ethernet card), logical address (e.g. 132.146.29.11, a TCP/IP address for a workstation) or a personal address (mnorris@iee.org, to reach an individual).

Address mask	Also known as a subnet mask. It is used to identify which bits in an IP address correspond to the network address and which bits refer to a local terminal.
Address resolution	The conversion of an Internet address into its corresponding physical address (for instance a corresponding Ethernet address).
Agent	In systems and network management a term usually applied to a server specialised from performing management operations on the target system or network node.
Agent	A more recent use of the word (sometimes prefixed with the word Intelligent) used to describe a semi-autonomous program that roams through a computing network collecting and processing data on behalf of its originator, sending back results as necessary.
Algorithm	A group of defined rules or processes for solving a problem. This might be a mathematical procedure enabling a problem to be solved in a definitive number of steps. A precise set of instructions for carrying out some computation (e.g. the algorithm for calculating an employee's take-home pay).
Analogue transmission	Transmission of a continuously variable signal as opposed to discretely variable signal. Telephony networks have traditionally been analogue.
ANSA	Advanced Networked Systems Architecture, a research group established in Cambridge, UK in 1984 that has had a major influence on architectures for distributed systems.
API	Application Programming Interface—software designed to make a computer's facilities accessible to an application program. It is the definition of the facilities offered to a programmer. All operating systems have APIs. In a networking environment it is essential that various machines' APIs are compatible, otherwise programs would be exclusive to the machines on which they reside.
APPC	Advanced Program to Program Communication—an application program interface developed by IBM. Its original function was in mainframe environments enabling different programs on different machines to communicate. As the name suggests the two programs talk to each other as equals using APPC as an interface designed to ensure that different machines on the network talk to each other.
Applet	A small(ish) software component of little use on its own but which may be plugged in to form part of a larger application. Used with World Wide Web applications, Java and mobile code environments to provide downloadable components.
Application	A collection of software functions (and possibly components) recognised as delivering most of what is needed to support a particular activity. Applications can be hand-crafted pieces of software but are more often commercial products or assemblies of reusable, black box components. Editors, spreadsheets, and text formatters are common

examples of applications. Network applications include clients such as those for FTP, electronic mail and telnet.

Application Generators

High level languages that allow rapid generation of executable code, sometimes referred to as 4GLs, Focus being a typical example.

Architecture

A high level conceptual representation showing how systems and components in a domain relate to one another and may be assembled into more complex systems. Any given domain may have a number of different architectures representing different viewpoints. When applied to computer and communication systems, it denotes the logical structure or organisation of the system and defines its functions, interfaces, data and procedures. In practice, architecture is not one thing but a set of views used to control or understand complex systems. A loose definition is that it is a set of components and some rules for assembling them.

Architecture style

A set of components, topological layout, set of interaction mechanisms, environment and possibly technology (e.g. CORBA).

ARP

Address Resolution Protocol is a networking protocol that provides a method for dynamically binding a high level IP address to a low level physical hardware address. This means, for instance, finding a host's Ethernet address from its Internet address. ARP is defined in RFC 826.

Asynchronous

An arrangement where there is no correlation between system time and the data that is exchanged, carried or transmitted over the system. For instance, an asynchronous protocol sends and receives data whenever it wants—there is no link to a master clock. The penalty for this freedom is that extra information has to be added to announce the start and stop of a communication.

Asynch data transmission

A data transmission in which receiver and transmission and transmitter clocks are not synchronised. Each character (word/data block) is preceded by a start by and terminated by one or more stop bits, which are used at the receiver for synchronisation.

ATM

Asynchronous Transfer Mode is a standard for high-speed fixed size packet communications. It provides a basis for multi-service networks—those capable of carrying voice, video, text etc.

Automata

A machine (such as a computer) that operates to a fixed set of instructions. Typically, automata do something, and then wait for an external stimulus before doing something else. The way in which automata interact is the subject of a well developed branch of mathematics known as automata theory.

Automation

Systems that can operate with little or no human intervention. It is easiest to automate simple mechanical processes, hardest to automate those tasks needing common sense, creative ability, judgement or initiative in unprecedented situations.

B

B Channel

The ISDN term used to describe the standard 64 kbit/s communications channel.

Bandwidth

The difference between the highest and lowest sinusoidal frequency

	signals that can be transmitted by a communications channel, it determines maximum information carrying capacity of the channel
Basic rate interface	An ISDN term that describes the two interfaces, 64 kbit/s transmission links and a 16 kbit/s signalling channel, referred to as bearer links and the delta channel. *See also ISDN.*
Batch processing	In data processing or data communications, an operation where related items are grouped together and transmitted for common processing.
BBS	Bulletin Board System. A computer based meeting and announcement system that allows people to both post and view information. Often organised into user groups or centred around a particular subject.
BGP	Border Gateway Protocol. This is the protocol used in TCP/IP networks for routeing between different domains.
Binding	This process whereby a procedure call is linked with a procedure address or a client is linked to a server. In traditional programming languages, procedure calls are assigned an address when the program is compiled and linked. This is static binding. With late, or dynamic, binding, the communicating parties are matched at the time the program is executed.
Bits per second	The basic measurement for serial data transmission capacity, abbreviated to bps. Usually has some form of modifier: kbps is thousands of bits per second, Mbps is millions of bits per second. Typically, a domestic user will have an Internet line running at a few tens of kbps. Backbone links are usually 2 Mbps and more.
Blocking	A situation when a path or connection is not available because all of those available are busy. Blocking is a phenomenon of circuit switched networks, where the designer trades off concentration against throughput. Most public switched networks are designed with sufficient resources to allow users to gain access virtually all the time, without being blocked by other users.
Bridge	A device or technique used to match circuits, thereby minimising any transmission impairment. Most commonly used to connect two segments of a local area network together.
Browser	A program which allows a person to read hypertext information. The browser gives some means of viewing the contents of nodes and of navigating from one node to another. Netscape and Internet Explorer are browsers for the World Wide Web. They act as clients to the array of remote servers on which Web pages are hosted.
Bug	An error in a program or fault in equipment. Origin of the term is not universally agreed but popular belief is that the first use in a computing context can be attributed to Vice-Admiral Grace Murray Hopper of the US Navy. In the early days of valve-based electronic computing she found that an error was caused by a genuine bug—a moth fluttering around inside the machine.

C

| C | A widely-used programming language originally developed by Brian Kernighan and Dennis Ritchie at AT&T Bell Laboratories. C |

became most widely known as the language in which the UNIX operating system was written.

C++   A programming language based upon C but adding many enhancements particularly for object-orientated programming. It has probably now surpassed C in popularity and provides the basis for Java, the sub-set of C++ now widely used for building mobile code.

Caching   This is a process by which data requested by the operating system of a computer is retrieved from RAM instead of from a hard disk (or some other mass storage media). Caching algorithms will check if the requested data is in its 'cache' (or RAM). The significance of this is that RAM access is an order of magnitude faster than today's mass storage devices, so the more accesses to the cache, the faster overall system performance will be.

CCITT   Consultative Committee of the International Telegraph and Telephone. Until the early 1990s, a key standards making body for public network operators. Superseded by ITU/T.

CEN/CENELEC The two official European bodies responsible for standards setting, subsets of the members of the International Standards Organisation (ISO). The main thrust of their work is functional standards for OSI related technologies.

CEPT   The European Conference of Posts and Telecommunications—an association of European PTTs and network operators from 18 countries. It is the sister organisation to CEN/CENELEC.

CERN   The European laboratories for particle physics. Home of the HTML and HTTP concepts that underpin the popular Netscape and Internet Explorer browsers.

CGI   Common Gateway Interface. A protocol associated with file servers for the World Wide Web. CGI is the logical interface between an HTTP server and an application server. It allows information (e.g. records taken from a database) to be presented to a user in a standard format.

Client   Often synonymous with a PC. A client is an entity—for example a program, process or person—that is participating in an interaction with another entity and is taking the role of requesting (and receiving) the required service.

Client-Server The division of an application into (at least) two parts, where one acts as the 'client' (by requesting a service) and the other acts as the 'server' (by providing the service). The rationale behind client-server computing is to exploit the local desk top processing power leaving the server to govern the centrally held information.

CIR   Committed Information Rate. In Frame Relay, this is the guarantee level of throughput.

Circuit   An electrical path between two points generally made up with a number of discrete components.

Circuit switching The method of communications where a continuous path is first established by switching (making connections) and then using this path for the duration of the transmissions. Circuit switching is used in telephone networks and some newer digital data networks.

CMIP/CMIP Common Management Information Protocol/Service. A standard developed by the OSI to allow network elements to be remotely managed.

Code A computer program expressed in the machine language of the computer on which it will be executed, i.e. the executable form of a program. More generally, a program expressed in representation that requires only trivial changes to make it ready for execution.

COM Common Object Model (*see also DCOM*). The expansion of the Component Object Model (see next entry) to add support for distribution. COM was jointly developed by Microsoft and Digital (now Compaq).

COM Component Object Model. The non-distributed framework underlying Microsoft's OLE object technology.

Components Self-contained, recognisable entities that perform well understood functions and can be assembled via known interfaces with other components to build something more complex. Components are often reused and can be replaced with an identical functioning component without affecting overall operation.

Computer A piece of hardware that can store and execute instructions (i.e. interpret them and cause some action to occur).

Concurrency The case when two or more systems cooperate on a task in parallel. Concurrent operation can be efficient but is prone to undesirable states (such as deadlock, where all parties are waiting, or livelock, where there is a repeated sequence of activity with no exit defined).

Connection-Orientated This is the familiar form of communication on the telephone network. A call is initiated by setting up an end-to-end connection between participants and this connection is kept for the duration of the call. It may not be efficient in terms of network usage, but there are some assurances of delivery.

Connectionless This refers to a communication where two or more participants do not have a fixed path between them. Each of the packets that constitute the communication looks after its own routeing. This arrangement is subject to the vagaries of network availability but can be a very efficient overall way of using a network.

Configuration A collection of items that bear a particular relation to each other (e.g. the data configuration of a system in which classes of data and their relationships are defined).

Cookie A token of agreement between cooperating programs that is used to keep track of a transaction. At a more concrete level, a cookie is a fragment of code that holds some information about your local state—your phone number or home page reference, for instance. You probably have cookies that you do not know about. The Netscape and Explorer browsers both support them, with the cookie being presented to the server to control your dialogue.

CORBA Common Object Request Broker Architecture. Strictly, the name of a framework specification produced by the Object Management Group describing the main components of a distributed object environment. Informally used to denote any of a number of related specifications produced by the OMG.

CSMA/CD	Carrier Sense Multiple Access with Collision Detection—a method used in local area networks whereby a terminal station wishing to transmit *listens* and transmits only when the shared line is free. If two or more stations transmit at the same time, each backs off for a random time before re-transmission. Each station monitors its signal and if this is different from the one being transmitted, a collision is said to have occurred (collision detection). Each backs off and then tries again later.

D

Daemon	A program that lies dormant waking up at regular intervals or waiting for some predetermined condition to occur before performing its action. Supposedly an acronym rationalised from Disk And Execution MONitor.
Data	Usually taken to mean the same as information. Data is the raw input which, once interpreted and processed, can be used to provide information. For instance, a spreadsheet contains data and the fact that it shows your company to be in profit the information!
Database	A collection of interrelated data stored together with controlled redundancy to support one or more applications. On a network, data files are organised so that users can access a pool of relevant information.
Database server	The machine that controls access to the database using client/server architecture. The server part of the program is responsible for updating records, ensuring that multiple access is available to authorised users, protecting the data and communicating with other servers holding relevant data.
Datagram	A variety of data packet. A self-contained, independent entity of data carrying enough information to be routed from source to destination without reliance on earlier exchanges between the source and destination.
Deadlock	A condition where two or more processes are waiting for one of the others to do something. In the meantime, nothing happens. A condition (undesirable) that needs to be guarded against, especially in the design of databases.
DBMS	DataBase Management System. A set of software used to set up and maintain a database that will allow users to call up the records they require. In many cases, a DBMS will also offer report and application generating facilities.
DCE	The Distributed Computing Environment. A set of definitions and software components for distributed computing developed by the Open Software Foundation, an industry led consortium. It is primarily an remote procedure call with a set of integrated services, such as security, time and directory.
DCOM	Distributed Common Object Model. Microsoft's upgrade to its initial version of COM for a distributed environment.
Directory	A directory provides a means of translating from one form information to another (e.g. someone's name into their telephone number). In a distributed system directory services are a key com-

ponent. They often perform much the same function as a telephone directory—translating from a symbolic name to a network address. A well known example is DNS, which translates Internet names (mnorris@iee.org) into IP addresses (142.119.42.17).

Distributed computing	In a move away from having large centralised computers such as minicomputer and mainframes, and bring processing power to the desktop. Often used as a synonym for distributed processing.
Distributed database	A database that allows users to gain access to records, as though they were held locally, through a database server on each of the machines holding part of the database. Every database server needs to be able to communicate with all the others as well as being accessible to multiple users.
DLL	Dynamic Link Library. A set of software utilities that are bound with source code when it is compiled.
DNS	Domain Name Service. A general-purpose distributed, replicated, data query service used on Internet for translating hostnames into Internet addresses. e.g. taking a dot address such as jungle.pdq.com and returning the corresponding numerical addresses.
DPE	Distributed Processing Environment. One of the various standards in this area.
DoD	The US Department of Defence. Notable for their sponsorship of the network that was to become the Internet.
Domain	In the broad context, this is a well understood area of common interest within which common technical terms are understood and common components can be practically applied When applied to networked communication systems, a domain is part of a naming hierarchy. An Internet domain name consists, for example, of a sequence of names or other words separated by dots.

E

ECMA	European Computer Manufacturers Association—an association comprised of members from computer manufacturers in Europe, it produces its own standards and contributes to CCITT and ISO.
EJB	Enterprise JavaBeans. Components written in the Java programming language intended to be run within a server based environment (e.g. a WWW server or database). EJBs run within a 'container' on the server and appear as objects to the outside world. Clients locate the EJB via the Java Naming and Directory Interface (JNDI).
Electronic mail	Messages automatically passed from one computer user to another, often through computer networks and/or via modems over telephone lines.
Electronic mail address	The coding required to ensure that an electronic mail message reaches its specified destination. There are many formats of mail address, perhaps the best known being the dot address used for Internet mail, e.g. 'mnorris@iee.org'.
Encapsulation	The transparent enveloping of one protocol within another for the purposes of transport. Encapsulation is, along with tunnelling, a favourite method for supporting multiple protocols across linked networks.

Encryption	A means of turning plain text into cipher text, hence protecting content from eavesdroppers. There are many ways of doing this, public and private key encryption being the two main ones.
Enterprise	A term (usually used a descriptor for 'network' or 'computing') to denote the resources deployed to suit the operating needs of a particular organisation.
ESIOP	Environment-Specific Inter-ORB Protocol. A protocol defined by the OMG for communication between ORBs.
Ethernet	A Local Area Network (LAN) characterised by 10 Mbit/s transmission using the CSMA/CD (Collision Sense Multiple Access with Collision Detection) access method. Ethernet was originally developed by and is a registered trademark of Xerox Corporation.

F

FDDI	Fiber Distributed Data Interface. An American National Standards Institute (ANSI) LAN standard. It is intended to carry data between computers at speeds up to 100 Mbit/s via fibreoptic links. It uses a counter rotating token ring topology and is compatible with the first, physical, level of the ISO seven-layer model.
Federation	A union of otherwise largely independent systems to support some common purpose. Federated systems share some basic agreements or protocols to enable them to work together but are operated and managed autonomously.
File server	A machine in local area network dedicated to providing file and data storage to other machines in the network.
Firewall	In general, this refers to the part of a system designed to isolate it from the threat of external interference (both malicious and unintentional).
Firewall machine	This is a dedicated machine that usually sits between a public network and a private one (e.g. between an organisation's private network and the Internet). The machine has special security precautions loaded onto it and used to filter access to and from outside network connections and dial-in lines. The general idea is to protect the more loosely administered machines hidden behind the firewall from abuse.
Flow Control	In a packet-switched network, packets compete dynamically for the networks resources—storage, processing, transmission. Flow control is a mechanism for ensuring fairness and controlling congestion.
Front-end	The presentation element of client-server system. Usually a desktop PC running a graphical user interface.
FTAM	File Transfer, Access and Manipulation. A protocol entity forming part of the Application Layer enabling users to manage and access a distributed file system.
FTP	File Transfer Protocol. The Internet standard (as defined in the RFC series) high level protocol for transferring files from one computer to another. A widely used de facto standard (c.f. the sparingly used *de jure* standard FTP). Anonymous FTP is a common way of allowing limited access to publicly available files via an anonymous login.

G

G.703   The CCITT standard for the physical and logical transmissions over digital circuit. Specifications include the US 1.544 Mbit/s as well as the European 2.048 Mbit/s that use the CCITT recommended physical and electrical data interface.

Gateway   Hardware and software that connect incompatible networks, which enables data to be passed from one network to another. The gateway performs the necessary protocol conversions.

GIOP   General Inter-ORB Protocol. A protocol defined by the OMG for communication between ORBs.

GUI   Graphical User Interface. An interface that enables a user to interact with a computer using graphical images and a pointing device rather than a character-based display and keyboard. Such interfaces are also known as 'WIMP' interfaces—WIMP standing for Windows, Icons, Menus and Pointers. The most common pointing device is that electronic rodent—the mouse.

H

Hardware   The physical equipment in a computer system. It is usually contrasted with software.

Heritage system   A euphemism for an old or decrepit system—synonyms include Legacy system, Cherished system and Millstone.

Heterogeneous   Of mixed or different type.

Homogeneous   Of the same type.

Hostage data   Data that is generally useful but held by a system which makes external access to that data difficult or expensive.

HTTP   Hyper Text Transfer Protocol. The basic protocol underlying the World Wide Web (see WWW). It is a simple, stateless request response protocol. Its format and use is rather similar to SMTP. HTTP is defined as one of the Internet's RFC series, generated by the IAB.

HTML   Hyper Text Markup Language. HTML is the language used to describe the formatting in WWW documents. It is an SGML document type definition. It is described in the RFC series of standards.

I

IAB   Internet Activities Board. The influential panel that guides the technical standards adopted over the Internet. Responsible for the widely accepted TCP/IP family of protocols. More recently, the IAB have accepted SNMP as their approved network management protocol.

Idioms   Pattern for implementing a solution in a specific language with a certain style. Effectively an implementation pattern.

IDL   Interface Definition Language. A notation that allows programs to be written for distribution. An IDL compiler generates stubs that provide a uniform interface to remote resources. IDL is used in conjunction with remote procedure calls.

IEEE	The Institute of Electrical and Electronic Engineers. US based professional body covering network and computing engineering.
IEE	UK equivalent of the IEEE.
IEEE 802.3	The IEEE's specification for a physical cabling standard for LANs as well as the method of transmitting data and controlling access to the cable. It uses the CSMA/CD access method on a bus topology LAN and is operationally similar to Ethernet. *See also OSI.*
IEEE 802.4	A physical later standard that uses the token ring passing access method on a bus topology LAN.
IETF	Internet Engineering Task Force. The IETF is a large, open international community of network designers, operators, vendors and researchers whose purpose is to coordinate the operation, management and evolution of the Internet and to resolve short- and mid-range protocol and architectural issues. It is a major source of proposals for protocol standards which are submitted to the Internet Architecture Board (IAB) for final approval.
IIOP	Internet Inter-ORB Protocol. A protocol defined by the OMG for communication between ORBs.
Inheritance	In object orientation this is a relationship between classes, a sub-class inherits from its super-class. Inheritance is (along with encapsulation and polymorphism) a basic property of all objects.
Interface	The boundary between two things: typically two programs, two pieces of hardware, a computer and its user or a project manager and the customer. The channel for the two entities is a conduit through which information passes. Information can consist of data or commands. An application programming interface (API) defines the commands and data that when sent through the channel enable a software application to be controlled.
Internet	A concatenation of many individual TCP/IP sites into one single logical network all sharing a common addressing scheme and naming convention.
internet	With a lower case 'i', this term denotes any set of networks interconnected with routers.
Internet address	The 32-bit host address defined by the Internet Protocol (IP) in RFC 791. The Internet address is usually expressed in dot notation, e.g. 128.121.4.5. The address can be split into a network number (or network address) and a host number unique to each host on the network and sometimes also a subnet address.
	The dramatic growth in the number of Internet users over the last few years has led to a shortage of new addresses. This is one of the issues being addressed by the introduction of a new version of IP, IPv6.
Intranet	A closed internet system running within an organisation connected to the Internet, and protected from unauthorised access, via a 'firewall'. Increasingly business are using Intranets to provide employees with desktop access to key business systems.
Interoperate	The ability of computers from different vendors to work together using a common set of protocols. Suns, IBMs, Macs, PC etc. all work

	together allowing each to communicate with and use the resources of the other.
IP	The ubiquitous Internet Protocol, one of the key parts of the Internet. IP is a connectionless (i.e. each packet looks after its own delivery), switching protocol. It provides packet routeing, fragmentation and re-assembly to support the Transmission Control Protocol (TCP). IP is defined in RFC 791.
Ipv6	The proposed successor to IP. It is a longer address but still compatible with IP. The aims are to extend the available address space, improve security, support user authentication and cater for delay sensitive traffic.
ISO	Commonly believed to stand for International Standards Organisation. In fact ISO is not an abbreviation—it is intended to signify commonality (from Greek Iso = same). The ISO is responsible for many data communications standards. A well known standard produced by ISO is the seven-layer Open Systems Interconnection (OSI) model.
ISO 9001	The standard most often adopted to signify software quality. In reality it assures no more than a basic level of process control in software development. The focus on process control (documentation, procedures etc.) sometimes promotes bureaucracy in ISO 9001 accredited organisations.
Isochronous	Data transmission in which a transmitter uses a synchronous clock but the receiver does not. The received detects messages by start/stop bits as in asynchronous transmission.
ISP	Internet Service Provider. This is most people's first point of contact with the Internet. An ISP usually offers dial up access via SLIP or PPP connections to a server on the Internet. Most ISPs also offer their customers a range of client software that can be used on the net.

J

Java	A programming language and environment for developing mobile code applications. Java is a subset of the C++ language and is widely used to provide mobile code application for use over the Internet.
JavaBean	Components written in the Java programming language, originally intended to be delivered over the Internet and run on a desktop client PC. (*See also* EJB)
JNDI	Java Naming and Directory Interface (*See* EJB).
JVM	Java Virtual Machine. The ubiquitous engine for running java code, in essence a software CPU. The idea is that any computer can equip itself with a JVM, a small program that allows Java applets (which are widely available over the Internet) to be downloaded and used.

K

Kermit	A communications protocol developed to allow files to be transferred between otherwise incompatible computers. Generally regarded as a backstop: if all else fails Kermit will get the files from A to B!

Kernel	The level of an operating system that contains the system level commands—the functions hidden from the user. This program is always running on a processor.
Key	The record identifier used in many information retrieval systems (i.e. database keys). The term is also used for the secret codes used to encrypt and decrypt information transmitted over public networks.

L

LAN	Local Area Network. A data communications network used to inter-connect data terminal equipment distributed over a limited area.
Language	An agreed-upon set of symbols, rules for combining them and meanings attached to the symbols that are used to express something (e.g. the Pascal programming language, job-control language for an operating system and a graphical language for building models of a proposed piece of software).
Legacy system	A system which has been developed to satisfy a specific requirement and is, usually, difficult to re-configure substantially without major re-engineering.
Life-cycle	A defined set of stages through which a development passes over time—from requirements analysis to maintenance. Common examples are the waterfall (for sequential, staged developments) and the spiral (for iterative, incremental developments). Life-cycles do not map to reality too closely but do provide some basis for measurement and hence control.

M

MAC	Media Access Control. This controls access to the shared transmission medium by framing/deframing data units, error checking and pro-viding access rights. Conceptually, the MAC is part of data link control and is important in the operation of LANs.
Mainframe	A computer (usually a large one, and often manufactured by IBM) that provides a wide range of applications to connected terminals.
Messaging	Exchanging messages. Often but not limited to the context of electronic mail.
Message Passing	Communication through the exchange of messages. Although not a rigorously used term message passing systems usually have the connotation of real-time immediate message exchange.
Message Queuing	A message passing technology augmented by a store-and-forward capability.
Method	A way of doing something—a defined approach to achieving the various phases of the life-cycle. Methods are usually regarded as functionally similar to tools (e.g. a specific tool will support a particular method).
Methodology	Strictly, the science or study of methods. More frequently used as a more important sounding synonym for method, process or tech-nique.
MIB	Management Information Base. The data schema that defines

	information available in an SNMP enabled device. MIB (now at version 2) is defined in RFC 1213.
Mips	Millions of instructions per second—one measure of a computer's processing power is how many instructions per second it can handle.
Middleware	Software that mediates between an application program and an underlying set of utilities such as a database, a network or a server. It manages the interaction between disparate applications across the heterogeneous platform, masking diversity from the programmer. Object Request Brokers such as CORBA are an example of Middleware, as they manage communication between objects, irrespective of their location.
Mobile code	Programs capable of being run on many different systems (e.g. Java can run on any machine equipped with a Java Virtual Machine). Mobile code is 'write once, use anywhere' and gets around the need for porting work to be done every time the program encounters a different type of computer.
Model	An abstraction of reality that bears enough resemblance to the object of the model that we can answer some questions about the object by consulting the model.
Modelling	Simulation of a system by manipulating a number of interactive variables; can answer 'what if ...?' questions to predict the behaviour of the modelled system. A model of a system or sub-system is often called a prototype.
Modularisation	The splitting up of a software system into a number of sections (modules) to ease design, coding etc. Only works if the interfaces between the modules are clearly and accurately specified.
MOM	Message Oriented Middleware. A term used to describe commercial message passing and message queuing products.
MP	Abbreviation used for both Multi-processing and Message Passing.
Multiplexing	The sharing of common transmission medium for the simultaneous transmission of a number of independent information signals. See Frequency Division Multiplexing FDM and Time Division Multiplexing TDM.
Multi-processing	Running multiple processes or tasks simultaneously. This is possible when a machine or has more than one processor or processing is shared among a network of uni-processor machines. *See also* multi-tasking and multi-threading.
Multi-processor	A single computer having more than one processor and capable of executing more than one program at once.
Multi-tasking	Performing (or seeming to perform) more than one task at a time. Multi-tasking operating systems such as Windows, OS/2 or UNIX give the illusion to a user of running more than one program at once on a machine with a single processor. This is done by 'time-slicing' dividing a processor into a small chunks which are allocated in turn to competing tasks.
Multi-threading	Running multiple threads of execution within a single process. This is a lower level of granularity than multi-processing or multi-tasking.

Threads within a process share access to the process memory and other resources. Threads may be 'time-sliced' on a uni-processor system or executed in parallel on a multi-processor system.

N

Network A general term used to describe the inter-connection of computers and their peripheral devices by communications channels. For example Public Switched Telephone Network (PSTN), Packet Switched Data Network (PSDN), Local Area Network (LAN), Wide Area Network (WAN).

Network interface The circuitry that connects a node to the network, usually in the form of a card fitted into one of the expansion slots on the back of the machine; It works with the network software and operating system to transmit and receive messages over the network to other connected devices.

Network operating system A network operating system (NOS) extends some of the facilities of a local operating system across a LAN. It commonly provides facilities such as access to shared file storage and printers. Examples include Novell's NetWare and Microsoft's LAN Manager.

Network topology The geometry of the network relating to the way the nodes are interconnected.

NFS Network File System. A method, developed by Sun Microsystems, that allows computers to share files across a network as if they were local.

Non-proprietary Software and hardware that is not bound to one manufacturer's platform. Equipment that is designed to the specification that can accommodate other companies' products. The advantage of non-proprietary equipment is that a user has more freedom of choice and a larger scope. The disadvantage is when it does not work, you may be on your own.

NNTP Network News Transfer Protocol. A protocol defined in RFC 977 for the distribution, inquiry, retrieval and posting of Usenet news articles over the Internet. It is designed to be used between a news reader client and a news server.

O

Object An abstract, encapsulated entity which provides a well defined service via a well defined interface. An object belongs to a particular class which defines its type and properties. One object can inherit properties from another and objects can evolve to do specific tasks.

Object orientation A philosophy that breaks a problem into a number of cooperating objects. Object-orientated design is becoming increasingly popular in both software engineering and related domains, for example in the specification of component based systems.

Object program The translated versions of a program that has been assembled or compiled. Nothing to do with object orientation!

ODP Open Distributed Processing. One of a number of organisations (most of which have the word 'open' in their title) which provide standards and/or components that allow computers from different vendors to interwork.

OLE
: Object Linking and Embedding. Microsoft's proprietary object component technology. Often compared to CORBA.

OMA
: Object Management Architecture (OMG).

OMG
: Object Management Group. An industry consortium responsible for the CORBA specifications.

OMNI
: Open Management Interoperability. An ISO based network management standards body. Responsible for OMNIPoint, which includes the CMIS and CMIP standards for connection between network elements and network management systems.

Open system
: A much abused term! The usual meaning of an open system is one built to conform published, standard specifications or interfaces, for example POSIX. Openness is rather like beauty in that it is often in the eye of the beholder.

Operating system
: Software such as VME, MVS, OS/2, Windows, VMS, MS-DOS or UNIX that manages the computer's hardware and software. Unless it intentionally hands over to another program, an operating system runs programs and controls system resources and peripheral devices.

OSF
: Open Systems Foundation. An organisation that provides generic UNIX based software components.

OSI
: Open Systems Interconnection. A model to support the interworking of telecommunications systems. The ISO Reference Model consisting of seven protocol layers. These are the application, presentation, session, transport, network, link and physical layers.

The concept of the protocols is to provide manufacturers and suppliers of communications equipment with a standard that will provide reliable communications across a broad range of equipment types. Also more broadly applied to a range of related computing and network standards.

OSPF
: Open Shortest Path First. A routeing protocol for TCP/IP networks.

OSS
: Operational Support Systems. These are all of the behind-the-scenes systems that allow a service to operate reliably and profitably. The usual spread of OSS includes billing, fault and problem management, provisioning, network and service management, customer handling and management information.

Overloading
: A term used in object-orientated software development to describe the use of one identifier that serves more than one purpose.

P

Packet
: A unit of data sent across a network. The basis for all of the modern data communication networks a common format for communications between computers.

Packet switching
: The mode of operation in a data communications network whereby messages to be transmitted are first transformed into a number of smaller self-contained message units known as packets. Packets are stored at intermediate network nodes (packet-switched exchanges)

and are reassembled unto a complete message at the destination. The ITU-T recommendation for packet-switching is X.25.

Parallel processing	Performing more than one process in parallel. Usually associated with computer-intensive tasks which can be split up into a large number of small chunks which can be processed independently on an array of relatively inexpensive machines. Many engineering and scientific problems can be solved in this way. It is also frequently used in high quality computer graphics.
Parameter	A variable whose value may change the operation but not the structure of some activity (e.g. an important parameter in the productivity of a program is the language used). Also commonly used to describe the inputs to and outputs from functions in programming languages. In this context they may also be known as 'arguments'.
Pattern	An approach to solving a given type of problem by using analogy/comparison with established or existing solutions of a similar type (and progressing through the application of templates or reference designs).
Peer to peer	Communications between two devices on an equal footing, as opposed to host/terminal, or master/slave. In peer to peer communications both machines have and use processing power.
Pipe	A feature of many operating systems, a pipe is a method used by processes to communicate with each other. When a program sends data to a pipe, it is transmitted directly to the other process without ever being written onto a file.
Polling	The process of interrogating terminals in a multi-point network in turn in a prearranged sequence by controlling the computer to determine whether the terminals are ready to transmit or receive. If any problems are detected within the normal sequence of operations, the polling sequence is temporarily interrupted while the terminal transmits or receives.
Polymorphism	A term used in object-orientated software development to describe an object that can be used in different ways, according to context.
PoP	Point of Presence. A site where there exists a collection of telecommunications equipment, usually modems, digital leased lines and multi-protocol routers. The PoP is put in place by an Internet Service Provider (ISP).

An Internet Service Provider may operate several PoPs distributed throughout their area of operation to increase the chance that their subscribers will be able to reach one with a low cost telecommunications circuit. The alternative is for them to use virtual PoPs (virtual points of presence) via some third party.

Port	n. A device which acts as an input/output connection. Serial communication ports or parallel printer ports are examples. Also, a pleasant after dinner drink.

v. To transport software from one system to another different system and make the necessary changes so that the software runs correctly, taking account of the specific calls and structures used on that system.

POSIX	Portable Operating System Interfaces. A set of international standards defining APIs based upon the UNIX operating system.
Process	The usual term for a program currently being run by an operating system. A process is assigned resources such as memory and processor time the operating system. The term 'task' is sometimes used as a synonym. *See also* multi-processing, multi-tasking and multi-threading.
Processor	That part of a computer capable of executing instructions. More generally, any active agent capable of carrying out a set of instructions (e.g. a transaction processor for modifying a database).
Proprietary	Any item of technology that is designed to work with only one manufacturer's equipment. The opposite of the principle behind Open Systems Interconnection (OSI).
Protocol	A set of rules and procedures that are used to formulate standards for information transfer between devices. Protocols can be low level (e.g. the order in which bits and bytes are sent across a wire) or high level (e.g. the way in which two programs transfer a file over the Internet).
Prototype	A scaled-down version of something, built before the complete item is built, in order to assess the feasibility or utility of the full version.

Q

Quality assessment	A systematic and independent examination to determine whether quality activities and related results comply with planned arrangements and whether these arrangements are implemented effectively and are suitable to achieve objectives.
Quality of Service	Measure of the perceived quality of a service. Usually based on tangible metrics such as time to fix a fault, average delay, loss percentages, system reliability etc.
Quality system	The organisational structure, responsibilities, procedures, processes and resources for implementing quality management.
Quality system standards	A quality system standard is a document specifying the elements of a quality system. The ISO 9001 standard (which is generally used to control software development) is a widely known and used quality standard.
Queuing	When a frame or packet is to be transmitted on a link, it may have to wait because another frame is being processed in front of it. The frame is placed in a buffer until the transmitter is free. Hence queuing systems (i.e. packet-switched systems) require buffers (matched to load and capacity) and introduce delay (as opposed to circuit-switching systems, which block).

R

RARP	Reverse Address Resolution Protocol which provides the reverse function of ARP. RARP maps a hardware address to an Internet address.
Remote Procedure Call	An RPC provides a distributed programming mechanism where an apparently local procedure call in a client causes an implementation of the procedure provided by a server to be invoked.

Repository | A data store holding (or pointing to) software and systems entities that designers and developers could reuse in the process of delivering new 'systems solutions'. The repository provides services to manage the creation, use, versions, maintenance, translation and viewing of these entities.

Recursion | *See* Recursion

Resolve | Translate an Internet name into its equivalent IP address or other DNS information.

Reuse | The process of creating software systems using existing artefacts rather than starting completely from scratch. Code, components, designs, architectures, operating systems, patterns etc. are all examples of artefacts that can be re-used. Also methods and techniques to enhance the reusability of software.

RFC | Request for Comment. A long-established series of Internet 'standards' documents widely followed by commercial software developers. As well as defining common Internet protocols RFCs often provide the implementation detail to supplement the more general guidance of ISO and other formal standards. The main vehicle for the publication of Internet standards, such as SNMP.

RIP | Routing Information Protocol, a standard gateway protocol defined in RFC 1388 that uses message broadcasts to a destination based on hop count.

RMON | Remote Monitoring management information database. Developed by the IETF, this extension of SNMP MIB 2 provides a means for tracking, storing and analysing remote network management information.

Routers | A router operates at level 3 of the OSI model. Routers are protocol specific and act on routeing information carried out by the communications protocol in the network later. A router is able to use the information it has obtained about the network topology and can choose the best route for a packet to follow. Routers are independent of the physical level (layer 1) and can be used to link a number of different network types together.

Routeing | The selection of a communications path for the transmission of information from source to destination.

S

Screen Scraping | A method of accessing a server where the client presents itself as being a direct interface to a human user. The client 'reads' information from the 'screen' presented by the server and 'sends' information as 'keystrokes' from the pretend user.

Server | An object which is participating in an interaction with another object, and is taking the role of providing the required service.

Service | A piece of work done; a facility provided. In the context of components, a service is the process and interface through which the predefined function of a component is accessed. Typically services are used to 'wrap' legacy systems and enable imbedded legacy functions to be used as if they were stand-alone components.

Session	The connection of two nodes on a network for the exchange of data —any live link between any two data devices.
SGML	Standard Graphical Markup Language. An international standard encoding scheme for linked textual information. HTML is a subset.
Signalling	The passing of information and instructions from one point to another for the setting up or supervision of a telephone call or message transmission.
Silver Bullet	The ultimate cure all for all of the ills of the software industry—the magic panacea. Over the years there have been many silver bullets that promised to remove all of the pain: object orientation and Java are probably the most recent saviours. In reality, there is no magic to make inherent complexity easily controllable, as explained in Fred Brooks' seminal paper 'No silver bullet'.
SNA	Systems Network Architecture—an IBM layered communications protocol for sending data between IBM hardware and software.
SNMP	Simple Network Management Protocol—consists of three parts: Structure of Management Information (SMI), Management Information Base (MIB) and the protocol itself. The SMI and MIB define and store the set of managed entities, SNMP transports information to and from these entities.
SMTP	Simple mail transfer protocol. The Internet standard for the transfer of mail messages from one processor to another. The protocol details the format and control of messages.
Socket	A mechanism for creating a virtual connection between processes. At the simplest level, an application opens the socket, specifies the required service, binds the socket to its destination and then sends or receives data.
Software Glue	Common term for the software that binds together applications, operating systems and other elements. The closest practical instantiation of software glue is probably the middleware developed for distributed systems (DCE, CORBA etc.) under the auspices of the OMG and others.
SOM	This is IBM's object orientated development environment that allows users to put together class libraries and programs. Associated with SOM is an OMG CORBA conformant object request broker (known as DSOM) for building distributed applications. *See also* COM, the Microsoft equivalent of SOM.
Sonet	A synchronous optical transmission protocol (UK equivalent is SDH). Sonet is intended to be able to add and drop lower bit rate signals from the higher bit rate signal without needing de-multiplexing. The standard defines a set of transmission rates, signals and interfaces for fibreoptic transmission.
SQL	Structured Query Language. A widely used means of accessing the information held in a database. SQL enables a user to build reports on the data held.
Stateful	When applied to a server, this term implies that it maintains knowledge about and context for its clients between requests.

Stateless	When applied to a server, this term implies that it maintains no knowledge about its clients between requests—each request is handled in isolation from all preceding and following requests.
Stove pipe	An independent set of systems supporting a new requirement represented by a vertical stack of functionality in a layered architecture with few interconnections with other systems in the architecture. Contrasts with using existing systems and components in the architecture.
Synchronisation	The actions of maintaining the correct timing sequences for the operation of a system.
Synchronous transmission	Transmission between terminals where data is normally transmitted in blocks of binary digit streams and transmitter and receiver clocks are maintained in synchronism.
Syntax	The set of rules for combining the elements of a language (e.g. words) into permitted constructions (e.g. phrases and sentences). The set of rules does not define meaning (this is covered by semantics), nor does it depend on the use made of the final construction.
System	A collection of independently useful objects which happen to have been developed at the same time.
	A collection of elements that work together, forming a coherent whole (e.g. a computer system consisting of processors, printers, disks etc.)
System Assembly	Another name for system integration.
System design	The process of establishing the overall (logical and physical) architecture of a system and then showing how real components are used to realise it.
System Integration	The process of bringing together all of the components that form a system with the aim of showing that the assembly of parts operates as expected. This usually includes the construction of components to carry out missing or forgotten functions and glue to interconnect all of the components.

T

TA	Terminal Adapter. A piece of equipment used with an ISDN connection to allow existing terminals to hook up. The equivalent of a modem.
TCP	Transmission Control Protocol. The most common transport layer protocol used on Ethernet and the Internet. It was developed by DARPA. TCP is built on top of Internet Protocol (IP) and the two are nearly always seen in combination as TCP/IP (which implies TCP running on top of IP). The TCP element adds reliable communication, flow-control, multiplexing and connection-orientated communication to the basic IP transport. It is defined in RFC 793.
TCP/IP	Transmission Control Protocol/Internet Protocol. The set of data communication standards adopted, initially on the Internet, for interconnection of dissimilar networks and computing systems.
Telnet	An TCP/IP based application that allows connection to a remote computer.

Throughput	A way of measuring the speed at which a system, computer or link can accept, handle and output information.
Topology	A description of the shape of a network, for example star, bus and ring. It can also be a template or pattern for the possible logical connections onto a network.
TP	Transaction Processing. Concerned with controlling the rate of enquiries to a database. Specialist software—known as a TP monitor—allows a potential bottleneck to be managed
Trading	Matching requests for services to appropriate suppliers of those services, based on some constraints.
Transaction	A single, atomic unit of processing. Usually a single, small 'parcel' of work which should either succeed entirely or fail entirely.
Transaction Processing	Originally a term that mainly is applied to technology concerned with controlling the rate of enquiries to a database. Specialist software—known as a TP monitor—allowed potential bottlenecks to be managed.
Transparency	Distribution transparencies provide the ability for some of the distributed aspects of a system to be hidden from users. For example, location transparency may allow a user to access remote resources in exactly the same way as local ones.
Tunnelling	Usually refers to a situation where a public network is used to connect two private domains so that the privacy of the overall link is maintained. To all intents and purposes, the same as encapsulation.

U, V

UML	Unified Modelling Language. The object-orientated notation adopted by the OMG and devised by Messrs Booch, Rumbaugh and Jacobson. UML unifies several of the main (and formerly disparate flavours) of object-orientated notation.
UNO	Universal Network Objects. A standard defined by the OMG for communication between ORBs.
URL	Uniform Resource Locator. Essentially, this is the form of address for reaching pages on the World Wide Web. A typical URL takes the form http://www.interesting.com/
Value Chain	A process description of a business's key activities showing where business functions add value to the products or services the business provides.
Vendor independent	Hardware or software that will work with hardware and software manufactured by different vendors—the opposite of proprietary.
Virtual circuit	A logical connection across a network (e.g. the transmission path through an X.25 packet-switched data network established by exchange of set-up messages between source and destination terminals).
Virtual device	A module allocated to an application by an operating system or network operating system, instead of a real or physical one. The user can then use a computer facility (keyboard or memory or disk or port) as though it was really present. In fact, only the operating system has access to the real device.

Virtual machine A software program that provides an implementation of an abstract processing environment. It supplies an execution engine for other programs compiled into byte-code which it interprets.

Virus A program, passed over networks, that has potentially destructive effects once it arrives. Packages such as VirusGuard are in common use to prevent infection from hostile visitors. *See also* Worm.

VPN Virtual Private Network. A combination of public and private resources that has been combined to give the user a network that looks like a coherent resource, suited to their particular needs. To all intents and purposes, a VPN is an Enterprise Network.

W

W3 Common abbreviation for World Wide Web, the Internet based distributed information retrieval system that uses hypertext to link multimedia documents. This makes the relationship of information that is common between documents easily accessible and completely independent of physical location.

WWW is a client-server system. The client software takes the form of a 'browser' that allows the user easily to navigate the information on-line. Well known browsers are Netscape and Mosaic. A huge amount of information can be found on World Wide Web servers.

Window A flow control mechanism the size of which determines the number of packets that can be sent before an acknowledgement of receipt is needed, and before more can be transmitted.

Windows A way of displaying information on a screen so that users can do the equivalent of looking at several pieces of paper at once. Each window can be manipulated for closer examination or amendment. This technique allows the user to look at two files at once or even to run more than one program simultaneously. Also the generic name (though not a registered trademark) for Microsoft's family of operating—Windows 97®, Windows NT®.

Worm A computer program which replicates itself—a form of virus. The Internet worm was probably the most famous—it successfully, and accidentally, duplicated itself across the entire system.

X, Y, Z

X.25 A widely used standard protocol suite for packet-switching. that is also approved by the International Standards Organisation. X.25 is the dominant public data communications packet standard. Many X.25 products are available on the market and networks have been installed all over the world.

X.400 A store and forward Message Handling system (MHS) standard that allows for the electronic exchange of text as well as other electronic data such as graphics and fax. It enables suppliers to interworking between different electronic mail systems. X.400 has several protocols, defined to allow the reliable transfer of information between User Agents and Message Transfer Agents.

X.500	A directory services standard that permits applications such as electronic mail to access information, which can either be central or distributed.
XNS	Xerox Network Systems. A protocol designed for interconnecting PCs. Used (in several guises) by Novell, Ungermann-Bass, Banyan and other LAN suppliers.
XML	eXtensible Markup Language. Developed by the World Wide Web Consortium for describing and exchanging information between applications and environments that are based on different technologies and have different semantic representations. It uses the Internet HTTP (HyperText Transport Protocol) for data exchange, but has a higher level metadata representation.
X/Open	An industry standards consortium that develops detailed system specifications drawing on available standards. It has produced standards for a number of distributed computing technologies. X/Open also licenses the UNIX trademark and thereby brings focus to its various flavours (e.g. HP-UX, AIX from IBM, Solaris from SUN etc.).
Yahoo	Yet another hierarchically organised offering. One of the many search utilities that can be used to trawl and crawl the information held on World Wide Web.
Zip	A compression programme, from PKWare, to reduce files to be sent over a network to a more reasonable size. This was originally popularised on MS-DOS but has now spread to other operating systems.

Index